IMAGES
of America

MOUNT VERNON

On the Cover: Two unidentified drivers pose by their milk delivery trucks outside the Carnation Company, manufacturer of Carnation Evaporated Milk. In 1907, Mount Vernon was chosen as the location for the company's third condensery. The city also had a second condensery, the Mount Vernon Cream Company, established in 1906. These two large condenseries, along with a thriving dairy industry, earned Mount Vernon the nickname "Milk City." (Courtesy of John Kamb Jr..)

IMAGES
of America

MOUNT VERNON

Jessica Bylund and Kari Hock

ARCADIA
PUBLISHING

ISBN 978-0-7385-9607-5

Published by Arcadia Publishing
Charleston, South Carolina

Printed in the United States of America

Library of Congress Control Number: 2012940645

For all general information, please contact Arcadia Publishing:
Telephone 843-853-2070
Fax 843-853-0044
E-mail sales@arcadiapublishing.com
For customer service and orders:
Toll-Free 1-888-313-2665

Visit us on the Internet at www.arcadiapublishing.com

This book is dedicated to the memory of Dick Fallis and Roger Fox—two gentlemen whose love of local history inspired us throughout our journey.

Contents

ACKNOWLEDGMENTS

This book would not have been possible without the generosity of countless people. While every picture is credited, and we wholeheartedly thank each and every contributor, there are some special people we must acknowledge even further.

We would like to thank John Kamb Jr. for answering our countless questions and allowing us to access his collection, which happened to contain the perfect cover photograph. Joan Fox graciously admitted us into her home for great conversation, coffee, and access to her late husband's photographic treasures. She was always open to questions and even helped us with some detective work. Mark Iverson let us look through thousands of negatives from the late Bill Forman, where we discovered many previously unpreserved gems. Ed and Donna Marlow let us into their home and shared Ed's historical photograph collection. Thank you to Donald McMoran and the staff at the Washington State University (WSU) Skagit County Extension as well as former WSU extension agent Marvin Jarmin for providing photographs and answering all of our agricultural questions. Lisette Mast generously provided photographs and explained the tulip industry to us. Mike and Jeanne Youngquist provided historical photographs from their family and from the county auditor's office. We would like to thank former mayors Bud Norris and Skye Richendrfer, current mayor Jill Boudreau, and the rest of the mayor's office for being generous and supportive. Thanks to Dan Royal for sharing Dick Fallis's materials. For providing photographs, information, and enthusiasm, we would like to thank the following: the Basye family, Supt. Carl Bruner and the staff at the Mount Vernon School Administration, Gregg Davidson and Skagit Valley Hospital, Rev. Ronald Deegan and the staff at First Baptist Church, and the Skagit County Historical Society.

As for behind-the-scenes assistance, we would like to thank Louis Saekow for artistry and technical support. We would also like to thank Hollie Cook for her editing skills. And last but not least, none of this would have been possible without the love and support of our families. The biggest thanks, with all our love, to Erik, David, Audrey, Henry, Nathan, and Hunter!

INTRODUCTION

Mount Vernon, located in Skagit County, has a population approaching 33,000 residents. Adjacent to the Skagit River, the city encompasses approximately 11 square miles of land and 0.3 square miles of surface water. It is easy for the casual traveler to pass Mount Vernon and its iconic tulip-painted smokestack while speeding down Interstate 5—perhaps headed south to Seattle or north to Bellingham. But in the days when water and rail travel dominated the landscape, Mount Vernon was the flourishing hub of Skagit County. And even now, every April, those same overlooked freeway exits are clogged with tourists coming to view the famous flower fields of the valley during the annual Tulip Festival.

While meandering the streets of downtown Mount Vernon, it is possible to glean traces of the city's beginnings and clues into her past. First Street is lined with brick buildings from the early 1900s and features a statue of founding pioneer Jasper Gates. A memorial to the men who cleared the river of two giant logjams is located on the revetment. Historic photographs hang in the train station. Yet most of these glimpses are merely shadowed pieces of the unknown puzzle that is Mount Vernon's history. Unfortunately, not much has been written about the topic. Most publications about Skagit County focus on the older (and, as one historian wrote, "more fashionable") towns of La Connor and Anacortes. While those cities were vital to the founding of the area and have rich histories of their own, it is important to note that Mount Vernon became the county seat, which brought many professionals to the area, along with the loggers, farmers, and tradespeople, who helped create the thriving city. Although difficult to imagine now, long ago, Mount Vernon was the cultural hub for all the towns nearby; it was home to several theatres and even an 1892 Opera House.

The story of Mount Vernon begins long before the arrival of the pioneers. Prior to any human influence, the land itself had a tale to tell. Volcanic activity from Mount Baker began more than one million years ago. Thousands of years ago, great glaciers shaped the valley and transformed the landscape. The Skagit River altered the valley via erosion and flooding as it flowed from the Cascade Mountains out to Puget Sound. That same river was stocked with fish and helped sustain lush forests filled and densely populated with giant fir and cedar trees 20 feet in circumference and 200 feet tall.

The earliest inhabitants of the Skagit Valley were Indians who relied upon the natural abundance of the area. They survived by fishing, hunting, and gathering, while living in small groups. Permanent villages consisted of cedar-plank longhouses. The tribe known as the Upper Skagit consisted of 11 small bands of Indians. They lived close to the river and ranged from modern-day Mount Vernon all the way to the Cascade Mountains. The group that lived closest to the eventual location of Mount Vernon was called the Nookachamps.

Beginning in the 17th century, Spanish, English, and American explorers came into contact with Puget Sound tribes to the south and west of Mount Vernon. It was not until 1846 that the first white settlers began to trickle into the Skagit Valley. Predictably, several conflicts sprang up between land-hungry settlers and the Indians. In the 1850s, the clashes escalated, and the territory's governor, Isaac Stevens, drafted several peace treaties. The Point Elliott Treaty was signed on January 22, 1855, by approximately 80 Puget Sound tribal leaders, including representation from the Upper Skagit tribe. The treaty called for the end of tribal warfare, the surrender of Indian lands to the federal government, and the confinement of the tribes to several small reservations. In exchange, the Indians retained hunting and fishing rights and were promised schools and medical care. The Upper Skagits were deemed too scattered to warrant either status as a separate tribe or entitlement to their own land. Most were taken to the Tulalip and Lummi reservations, while some remained on traditional lands along the Skagit River.

Innocuous trappers or prospectors would travel in the area, but it was not until the 1870 arrival of the Northern Pacific Railroad surveyors and the subsequent influx of more settlers that the Upper Skagit Indians found themselves in conflict with new arrivals. They were enraged when settlers crossed over lands that were sacred burial grounds, but with so few members remaining, they were not able to launch any serious offensives against the newcomers. Some Indians even integrated into life with the white settlers. Unfortunately, the Upper Skagits began to dwindle even more due to diseases brought to the area by these new inhabitants.

It is around the time of the settlers and new arrivals that this pictorial history begins. The first settlers arrived in the 1860s; the town founders soon followed, laying out Mount Vernon, named after the home of George Washington, in 1877. Mount Vernon was officially incorporated in 1890 and was named county seat of the newly formed Skagit County in 1884. These events occurred in parallel with the development of modern photography. The first permanent photograph was produced in 1826 by the French inventor Joseph Nicéphore Niépce. The daguerreotype was developed in 1839, and in 1884, George Eastman invented film, which freed photographers from carrying boxes of plates and toxic chemicals. Photography became available to the masses in 1901 with the introduction of the Kodak Brownie. These developments and accomplishments made it seem possible to capture images for eternity. Unfortunately, most photographs, being susceptible to damage, do not survive. The preservation of visual history is paramount; pictures tossed in a box and forgotten could hold clues for future generations about their past.

The collection of photographs within this book is the result of countless meetings, phone calls, and independent research. This has not been an easy project. Mount Vernon does not have a museum or historical society of its own, nor has there been much focus given to Mount Vernon in books about Skagit County. This project has taken a full year in the trenches. Along the way, many amazing people have contributed both their time and their images. They have shared their love of history and of Mount Vernon to help this project. The ability to bring all these pieces together through so much hard work is a great gift to, but especially from, the community. There are more pieces to the puzzle, more pictures not yet discovered and more stories not yet recorded. Perhaps this compilation will serve as the inaugural survey of the captivating and delightful history of Mount Vernon, Washington.

One

Pioneers and Logjams
Early Years to 1880

When the first settlers began arriving in the 1860s, the Skagit River (the area's main mode of transportation) was blocked by two massive, ancient logjams. The first jam was half a mile long and 14 feet deep. To its north was an even more formidable logjam a mile and a half long. The logjams were so solidly packed that it was possible to cross them on foot; they had been there so long that trees up to 90 feet tall grew from them. Goods could not make it upriver beyond the blockage, and the wealth of the interior could not be rafted or barged downriver.

These daunting facts did not dissuade the first known settlers. The earliest on record were David E. Kimble, Jasper Gates, and Joseph F. Dwelley, who all settled between 1869 and 1870. Several more followed the next year, and by 1872, the first school was started in a building on the Kimble ranch. In 1876, Harrison Clothier, originally from Corinth, New York, took a winter job teaching 20 students in the school. Edward G. English, who had been a pupil of Clothier's in high school in Wisconsin during the mid-1860s, had also migrated west. The two men formed a partnership in March 1877 with plans to open a trading post. For $100, they purchased 10 acres of land (which encompassed four blocks on the waterfront) from Jasper Gates. They prepared a town plat and named their aspiring community Mount Vernon after the Virginia home of the nation's first president.

Meanwhile, a movement started to remove the logjams. Because of the immensity of the task, the settlers requested aid from the federal government in 1875. No aid came, however, and in 1876, a group of settlers took on the project themselves, hoping to profit by selling salvaged logs. It took six months to cut a 250-foot channel through the lower obstruction and an additional two years to breach the upper blockage. By the summer of 1879, the upper Skagit was open to navigation.

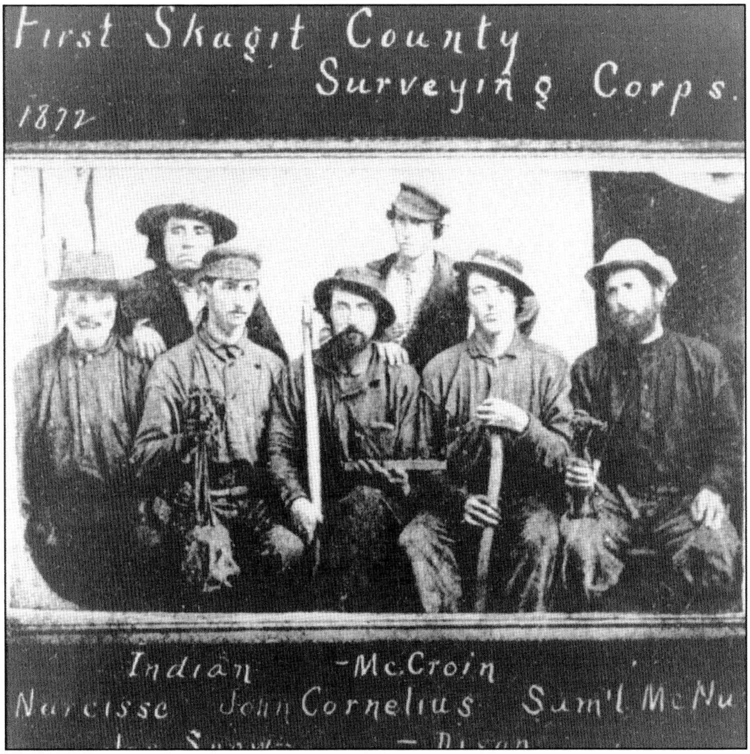

The Skagit County Survey Crew from 1872 (left) was led by John Cornelius, who obtained federal commissions to survey much of the surrounding area. A settler could not file an official homestead claim until after the survey was completed in the fall of 1872. Prior to that, settlers had to squat on a desired piece of property. The picture below shows a similar, unidentified survey crew south of Mount Vernon. (Left, courtesy of the Mount Vernon Mayor's Office; below, courtesy of the Roger Fox collection.)

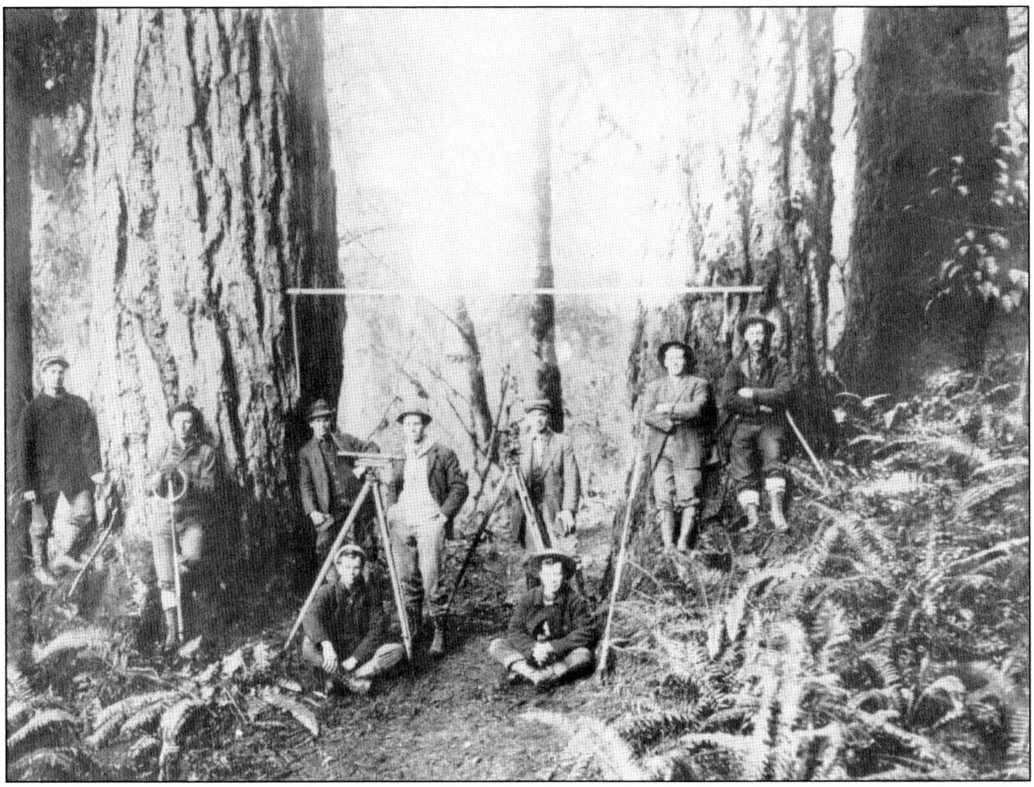

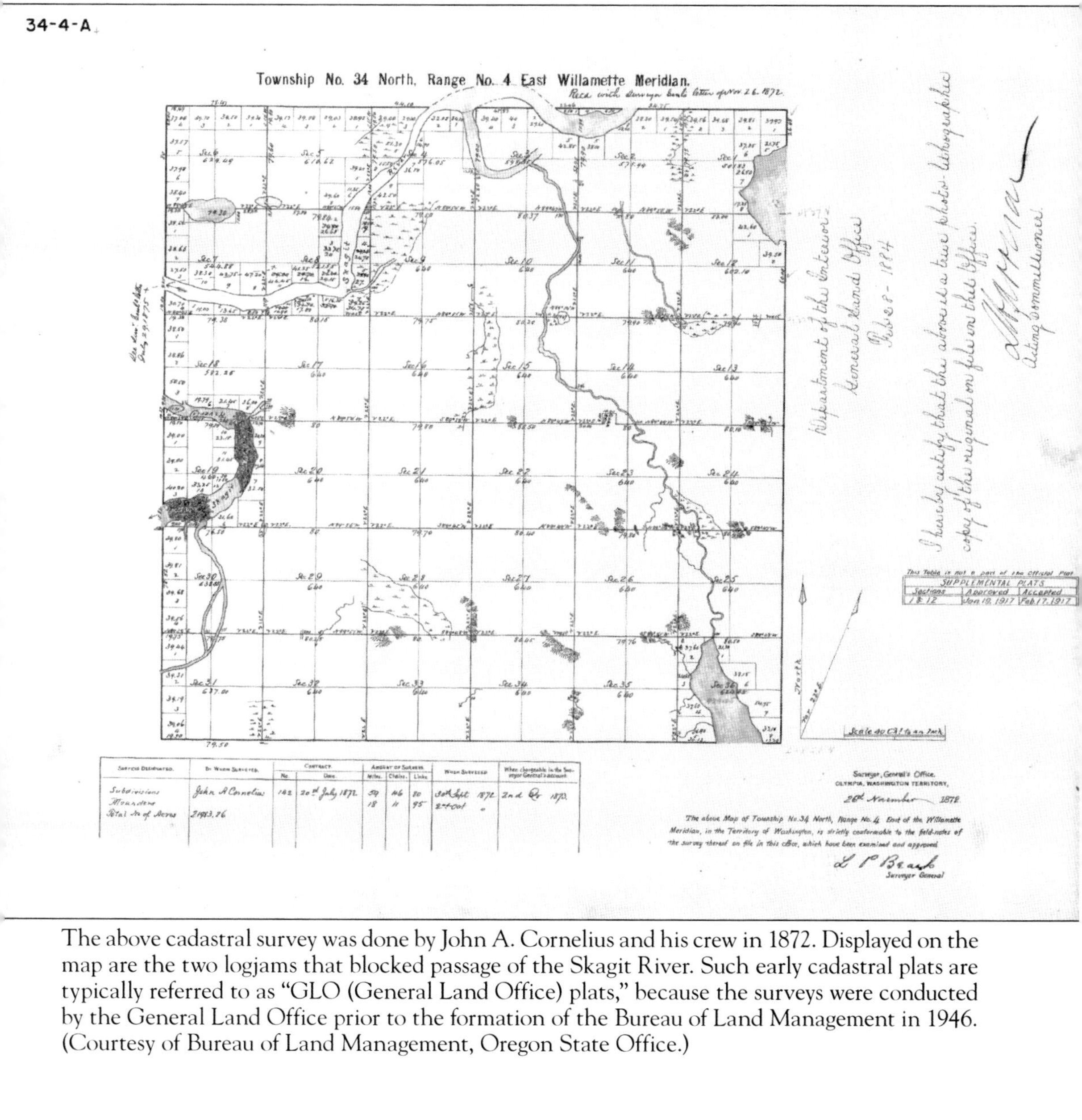

The above cadastral survey was done by John A. Cornelius and his crew in 1872. Displayed on the map are the two logjams that blocked passage of the Skagit River. Such early cadastral plats are typically referred to as "GLO (General Land Office) plats," because the surveys were conducted by the General Land Office prior to the formation of the Bureau of Land Management in 1946. (Courtesy of Bureau of Land Management, Oregon State Office.)

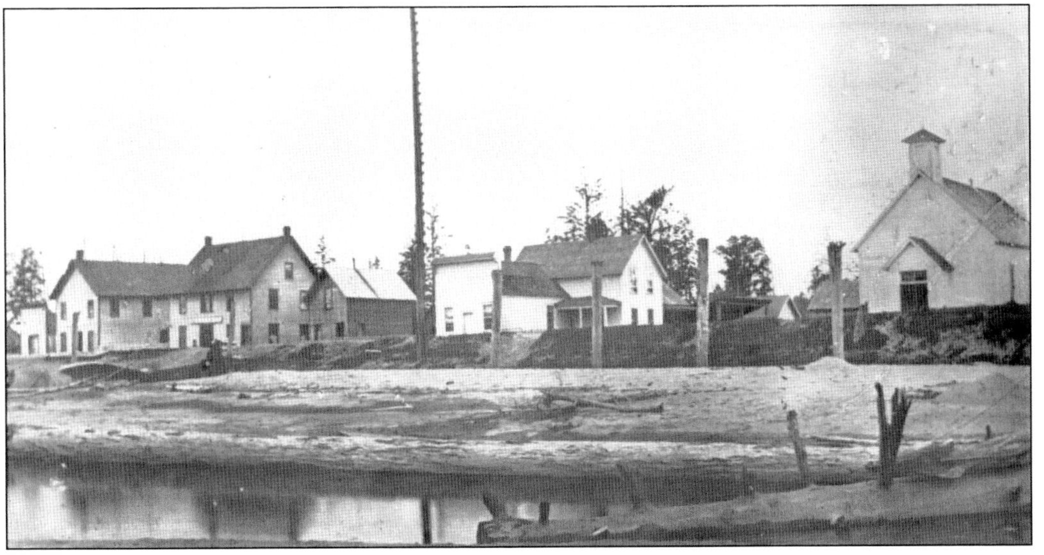

In 1869, John Barker opened a trading post at the south fork of the Skagit River on Fir Island that would evolve into Skagit City, the oldest of the river towns. A post office was established there in 1872, and by 1874, stern-wheelers traveled there on a monthly basis to bring supplies. Unfortunately, Skagit City slowly began declining once the logjams were cleared upriver. (Courtesy of the Skagit County Historical Society.)

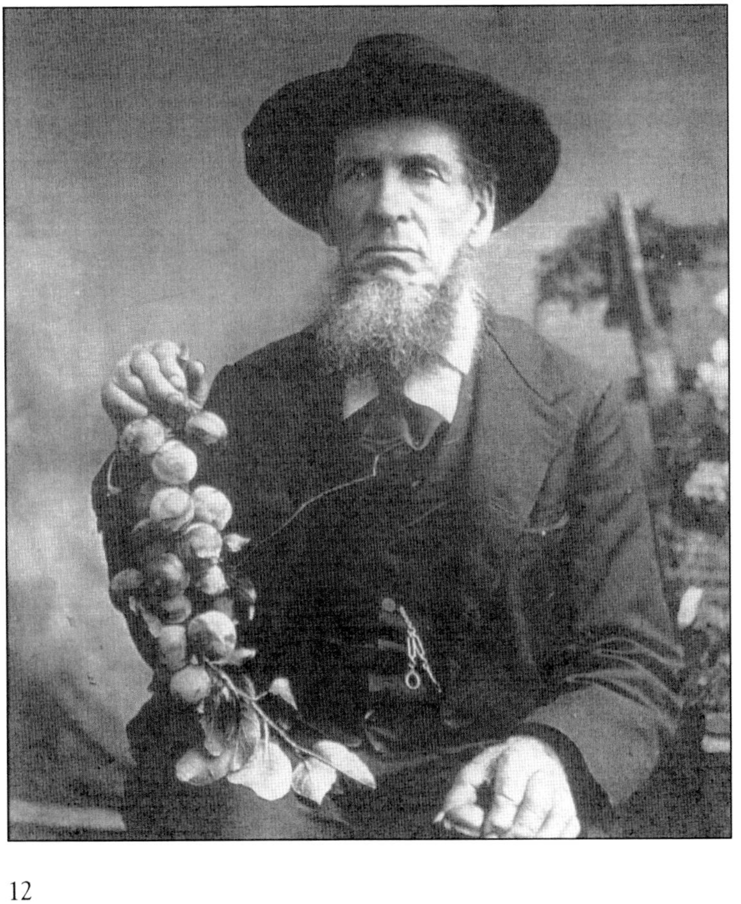

David Kimble moved from Illinois and staked a claim near the Skagit River on February 3, 1869, close to the southern end of the lower logjam. In February 1870, Kimble's family, including his second wife, Minerva Jane Bozarth, and others chartered the stern-wheeler *Linnie* for $50 to bring them and their household goods from Whidbey Island up the Skagit River. (Courtesy of Dan Royal.)

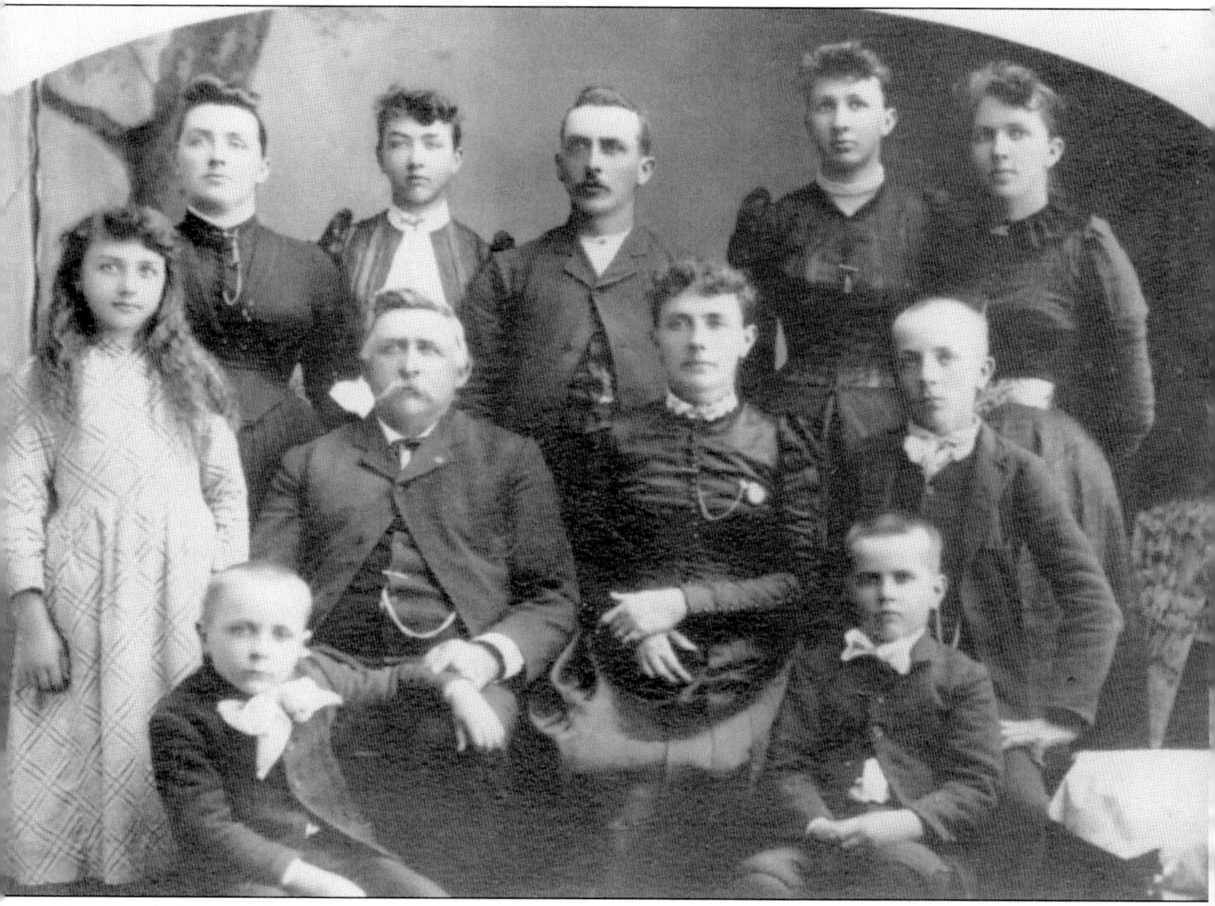

Pictured above is the Jasper Gates family c. 1900. Jasper and Clarinda C. Kimble Gates (married on October 6, 1860) are seated in the middle, surrounded by nine of their 11 children. Although not identified in the image, the children were, from oldest to youngest, Jasper N. (1863), Matilda (1863), Nancy Maria (1868), Mary (1873), Martha Jane (1875), Otto (1877), Clarinda Caroline (1880), William L. (1883), and Cleon Emmett (1885). Missing from the photograph are Nancy's twin sister, Ida Alis, who died in 1873, and Lawrence Gates, who was born and died in 1882. Jasper Gates, born in Missouri on April 9, 1840, was the 10th of 15 children. In 1869, he moved his family to the Washington Territory with two other related families. That same year, Gates and several other pioneers scouted the Skagit River. He eventually filed a homestead claim on land that would later be part of the city of Mount Vernon. Because Whatcom County did not have a land office, he had to travel all the way to Olympia to file his claim. (Courtesy of Mike Youngquist.)

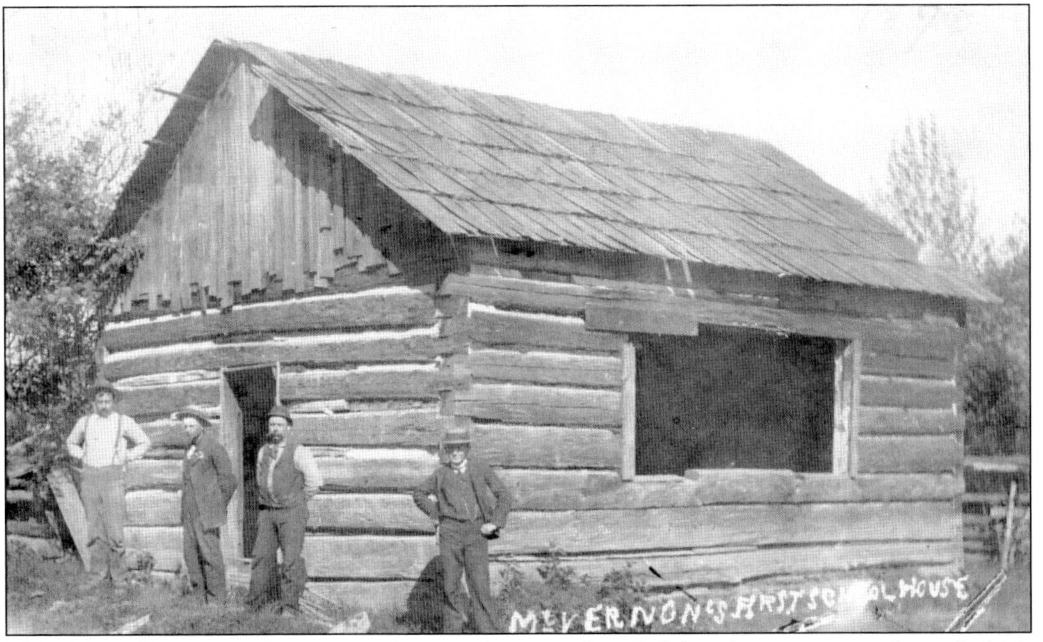

The men in this turn-of-the-century photograph are unidentified, but the structure in front of which they stand is known to be the outbuilding on David Kimble's property that was originally used as the first school in the future Mount Vernon area. (Courtesy of the Roger Fox collection.)

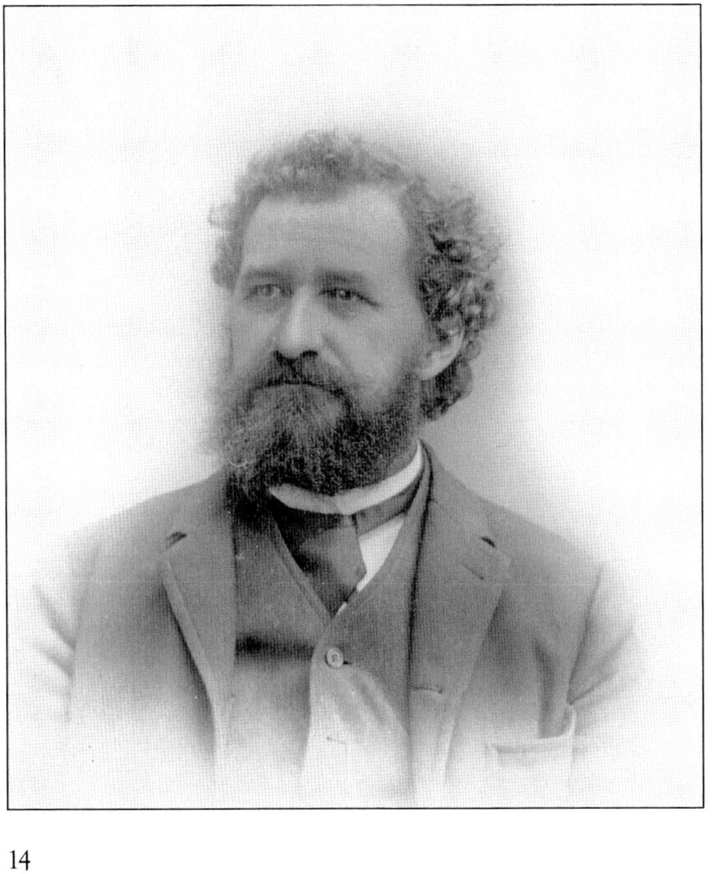

Harrison Clothier was born on July 9, 1840, in Corinth, New York. He was a teacher in New York, Wisconsin, and Minnesota before taking a winter teaching job at the homestead of David Kimble in 1876. Clothier forged a partnership in 1877 with a former pupil from Wisconsin, Edward G. English, to open a trading post on land they purchased from Jasper Gates. (Courtesy of the Skagit County Historical Society.)

Seen here in a picture from the late 1880s are, from left to right, (first row) James O'Loughlin (first sheriff of Skagit County) and an unidentified man; (second row) Harrison Clothier and Edward G. English. Born in Maine in 1850, English came to Skagit County in 1873. Besides founding Mount Vernon with partner Clothier, English founded the English Logging Company in 1902. (Courtesy of the Skagit County Historical Society.)

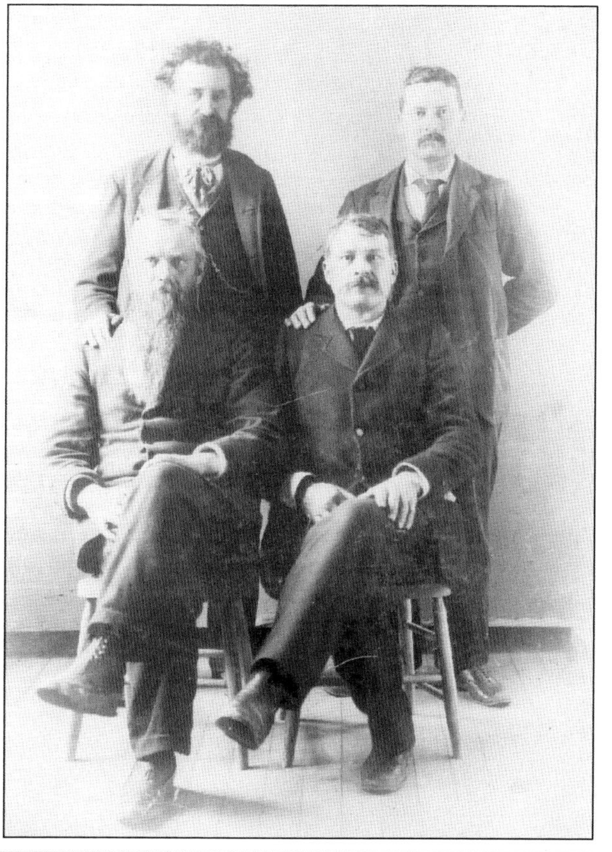

When settlers arrived in Mount Vernon, they encountered massive fir and cedar trees. Settlers cleared the trees, using the lumber for building or milling. Horses and oxen were used to transport large sections of timber to the river. This unidentified logging crew used a four-horse team to haul out part of a fir tree while logging above East Mount Vernon. (Courtesy of the Roger Fox collection.)

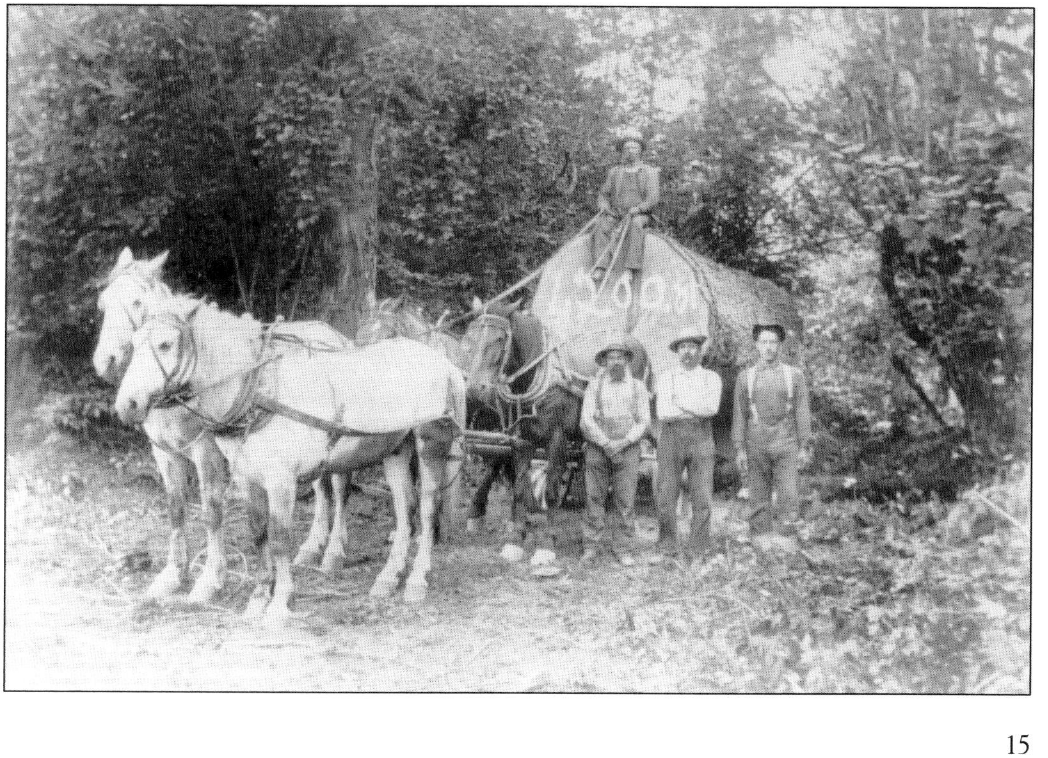

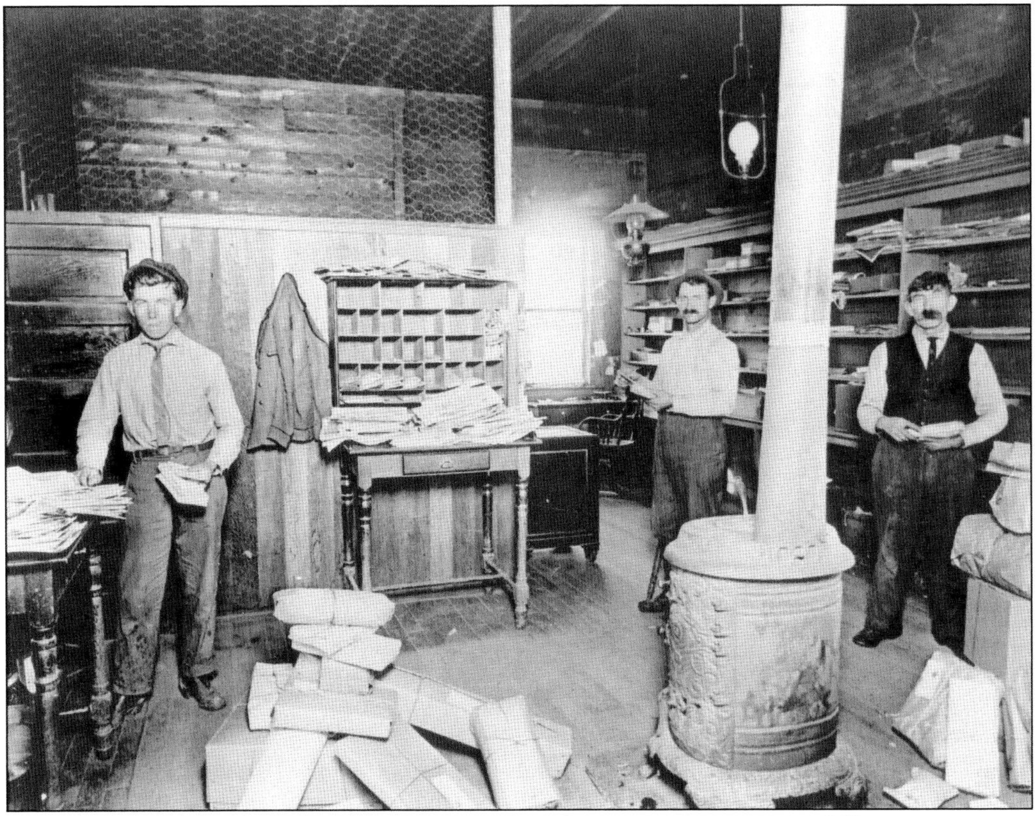

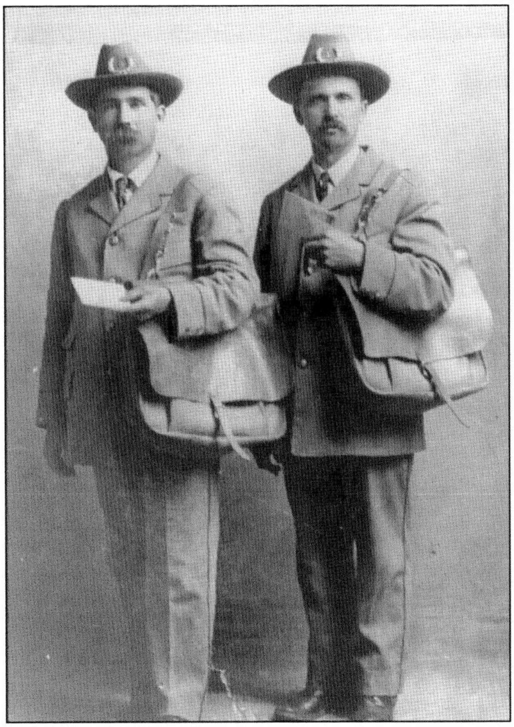

The Mount Vernon Post Office was authorized on June 29, 1877, with Clothier appointed postmaster. Because of the logjam, Jasper Gates was paid $1.50 to row his canoe to Skagit City to pick up the post from the Seattle-based *Fanny Lake* mail boat. That arrangement lasted until 1878, when the lower logjam was cleared enough to allow the steamer *Wenat* and, subsequently, other stern-wheelers to bring the mail directly to Mount Vernon. In 1881, English succeeded Clothier as postmaster. One of the early post office locations is pictured above. At left are two mail carriers from the turn of the century. (Above, courtesy of Ed Marlow; left, courtesy of the Roger Fox collection.)

Two

The Town that Should Not Have Been 1880–1890

Once the logjam was successfully cleared, prosperity flowed through the area. The ability to navigate the river to Mount Vernon and beyond made it possible to expand logging operations in the fertile forests of the region. In turn, cleared forests became productive farmlands that remain a staple of the area today. Additionally, prospectors poured through the town on their way upriver, hoping to find their fortunes at Ruby Creek, where gold had been discovered. Many miners arrived in Mount Vernon, via steamer, from Seattle; once in the town they would purchase supplies before continuing upriver in smaller vessels (typically canoes).

Since its founding, Mount Vernon had been part of Whatcom County. In November 1883, a push by some territorial legislators to form a new county out of what was Southern Whatcom resulted in the successful creation of Skagit County. The county seat was initially La Conner, but in November 1884, voters elected to move it to Mount Vernon. The decision was attributed to support from the logging, mining, and farming communities that had come to rely on Mount Vernon. The election results were disputed for many years, but the writing was on the wall, and many professionals, such as bankers and lawyers, began settling in the up-and-coming city.

In a few short years, Mount Vernon had transformed from a simple idea and a small settlement into a major city in the new Skagit County. The first school was built in 1881, followed by a larger school in 1884. That same year, the town saw the publication of its first newspaper, *Skagit News*. In 1888, the town's first sawmill was built on the riverfront. The first church was built in 1889.

Oddly, though, the citizens of Mount Vernon had never petitioned the territorial legislature for official incorporated status. Washington achieved statehood on November 11, 1889, and the following year, over 100 of Mount Vernon's nearly 1,000 citizens signed a petition asking for incorporation status. Mount Vernon was officially incorporated in 1890 and held its first municipal election on June 27.

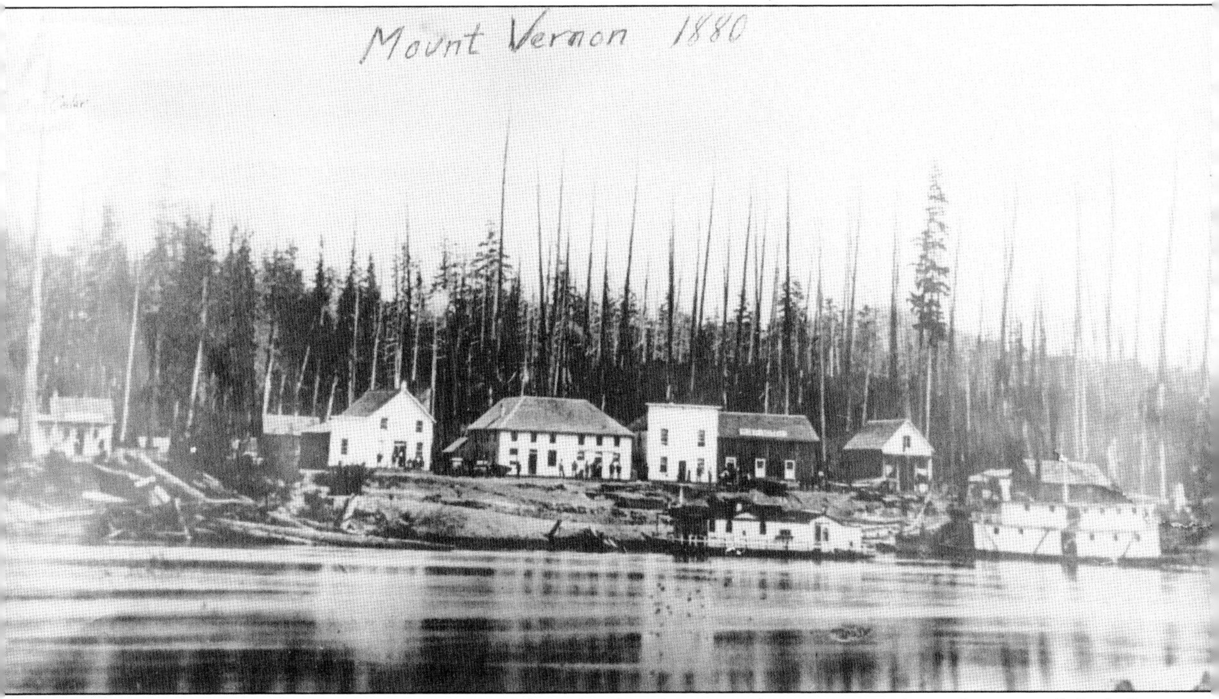

Mount Vernon is pictured in 1881, although the photograph appears to have been mislabeled. The buildings are, from left to right, the Bonanza Saloon, the Clothier & English Store (the first building in town, newcomers could spend the night in the loft upstairs; it later became known as Skagit Ned's), the Mount Vernon Hotel (the town's second hotel, built and run by George V. Brann and George Moran), the Ruby House (the town's first hotel, built and run by Michael McNamara), a drugstore owned by Dr. D.Y. Deere, and, finally, a logging camp. On the river can be seen a floating house (left) and the steamer *Glide* (right). Visible on the far left is the flagpole, which was created in 1877 when branches were cleared from a 147-foot-high cedar tree with a six-foot diameter. The flag measured 24-by-36-feet and was sewn by ladies of the community. Currently, the flag resides in a display case, funded by the Daughters of the American Revolution, at the county courthouse. (Courtesy of the Roger Fox collection.)

At right is a copy of a page from the official handwritten 1885 Washington Territory Census. It is interesting to note the details recorded about the Skagit County citizens. While there is not much variety in the profession column, and men seem to dominate the list, the nativity column is remarkably diverse. (Courtesy of the Skagit County Auditor's Office.)

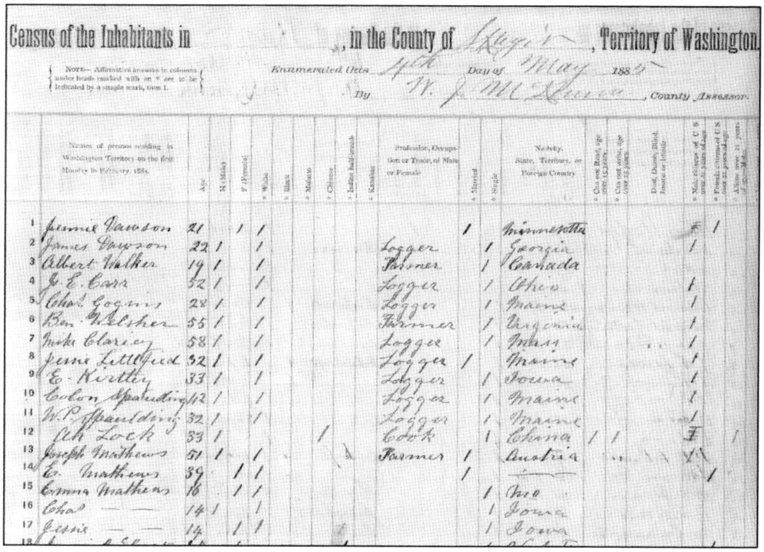

Census of the Inhabitants in , in the County of Skagit, Territory of Washington.

Enumerated this 4th Day of May 1885 By W. J. McAlison, County Assessor.

	Names of persons residing in Washington Territory on the first Monday in February, 1885	Age	M (Male)	F (Female)	White	Chinese	Profession, Occupation or Trade, of Male or Female	Married	Single	Nativity, State, Territory, or Foreign Country	Can not Read	Can not write	Male citizens of U.S.	Female citizens of U.S.	Aliens over 21 years
1	Jennie Dawson	21		1	1			1		Minnesotta				1	
2	James Dawson	22	1		1		Logger		1	Georgia			1		
3	Albert Walker	19	1		1		Farmer		1	Canada			1		
4	J. E. Carr	52	1		1		Logger		1	Ohio			1		
5	Chas. Goggins	28	1		1		Logger		1	Maine			1		
6	Ben Welsher	55	1		1		Farmer		1	Virginia			1		
7	Mike Clancy	58	1		1		Logger		1	Mass			1		
8	Jesse Littlefield	32	1		1		Logger	1		Maine			1		
9	E. Kirtley	33	1		1		Logger		1	Iowa			1		
10	Colon Spaulding	42	1		1		Logger		1	Maine			1		
11	W. P. Spaulding	32	1		1		Logger		1	Maine			1		
12	Ah Lock	33	1			1	Cook		1	China	1	1			1
13	Joseph Mathews	51	1		1		Farmer	1		Austria			1		
14	E. Mathews	39		1	1			1		—				1	
15	Emma Mathews	16		1	1				1	Mo.					
16	Chas —	14	1		1				1	Iowa					
17	Jessie —	14		1	1				1	Iowa					

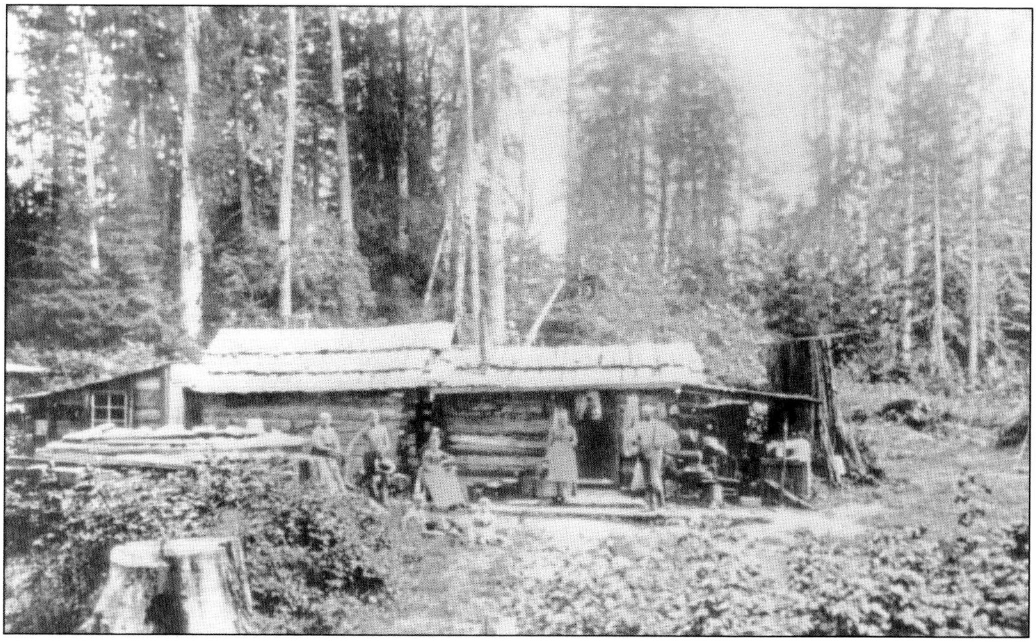

After land was logged, tremendous cedar and fir tree stumps were left behind. Homesteads on such land were coined "stump ranches." To clear fields for farming, oxen and horses were used to dig out and remove the stumps. Burning was another method of removal, but fire was a liability and eventually needed to be regulated in order to protect the surrounding forestland. (Courtesy of the Roger Fox collection.)

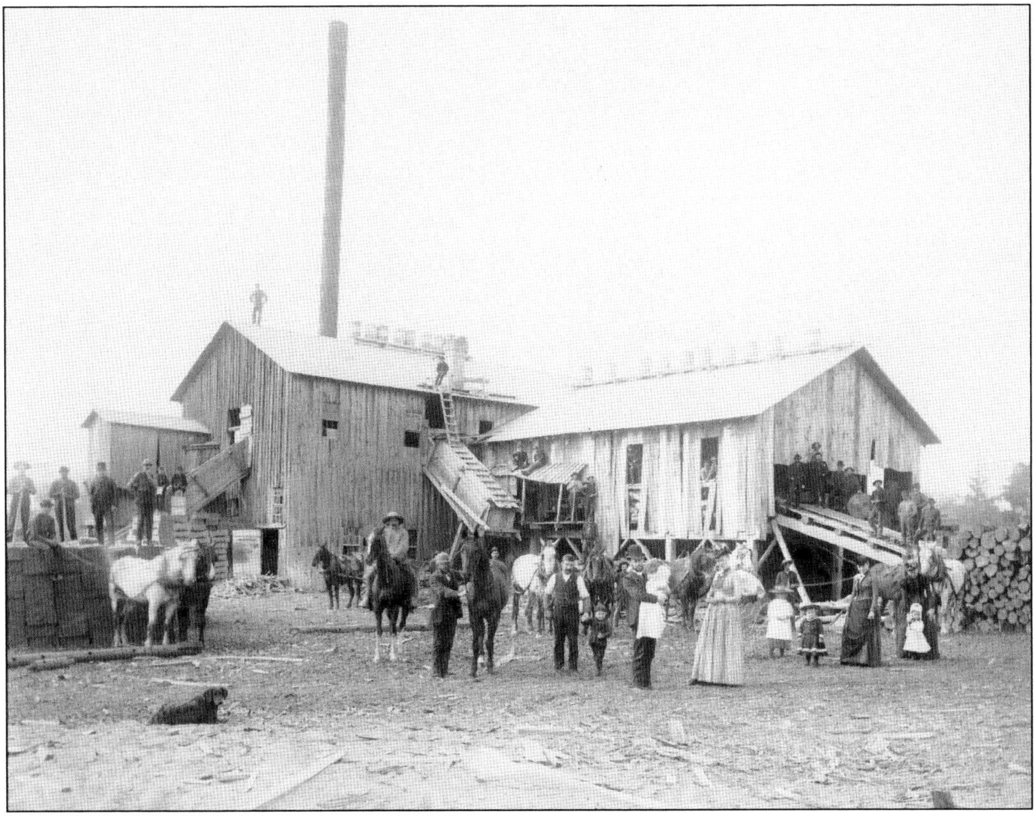

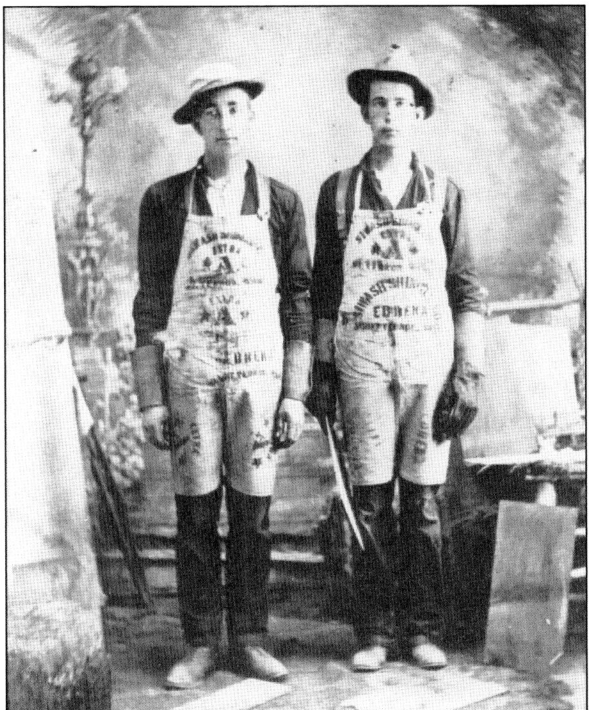

With a capacity of nearly 200,000 shingles per day, Siwash Shingle Mill was one of the largest shingle mills in the state. It employed many of Mount Vernon's citizens before burning down in 1909. There was little market for cedar until the late 1880s, when Mortimer Cook of Sedro (which became Sedro-Woolley in 1898) discovered that cedar shakes and shingles could be dried in kilns, making a lightweight product that could be economically shipped to growing markets across the country. The cedar was cut and woven into bundles for shipping. At left are two unidentified shingle weavers. The use of the term "weaver" for a shingle maker referred to the way they moved their hands with dexterity around the blades of their saws. (Above, courtesy of the Skagit County Historical Society; left, courtesy of the Roger Fox collection.)

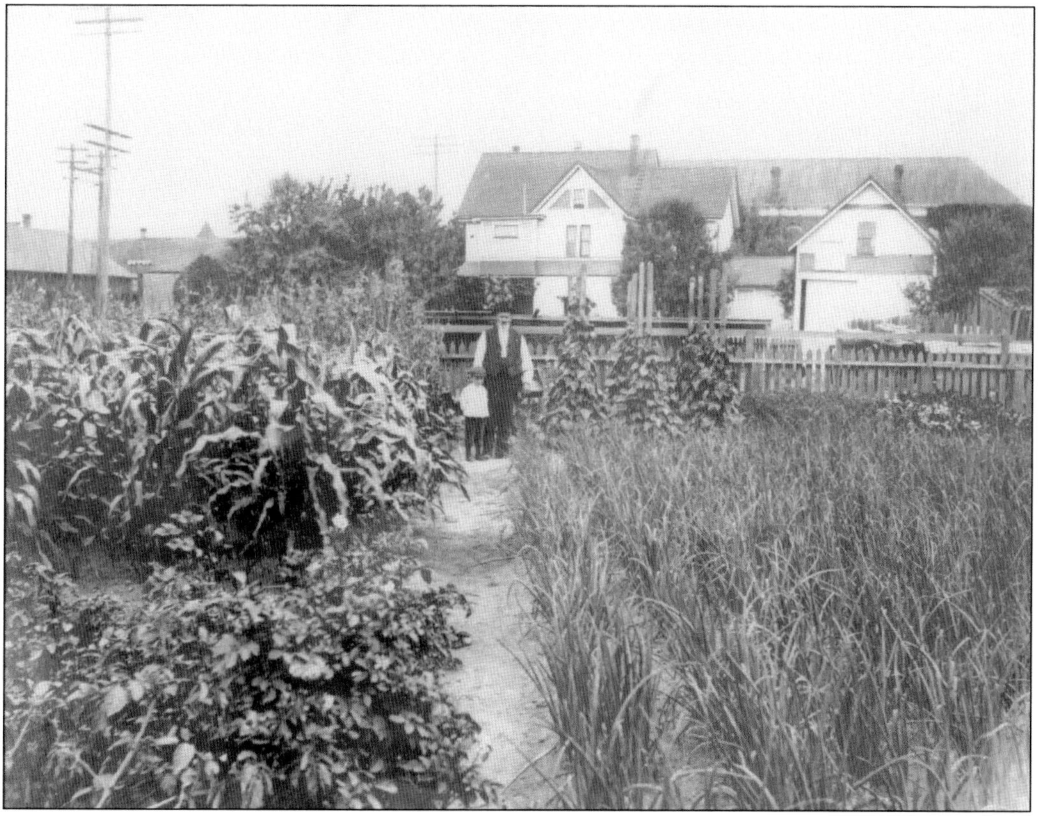

Pictured above (c. 1904) are Capt. David F. Decatur and his grandson Elden D. Pollock on Decatur's farm south of Kincaid Street. The house and barn are visible behind them, and the courthouse can be seen farther in the background. Decatur, a Massachusetts native, arrived in Mount Vernon in December 1887. That same year, the Skagit Sawmill & Manufacturing Company was incorporated by, among others, Clothier and English. It was one of the many investments they would make in the logging industry. The following year, Decatur built the first mill on the riverfront. It provided many benefits to the town, including employment for up to 45 men. The mill would have been similar to the Kellogg Lumber Mill, pictured below, which was located on Knapp Road east of Big Rock. (Both, courtesy of the Roger Fox collection.)

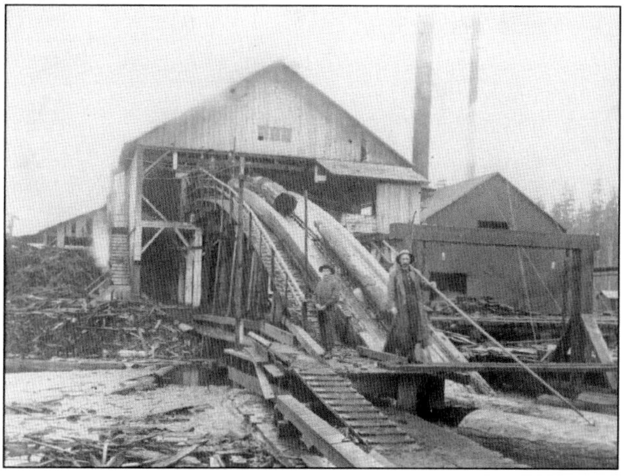

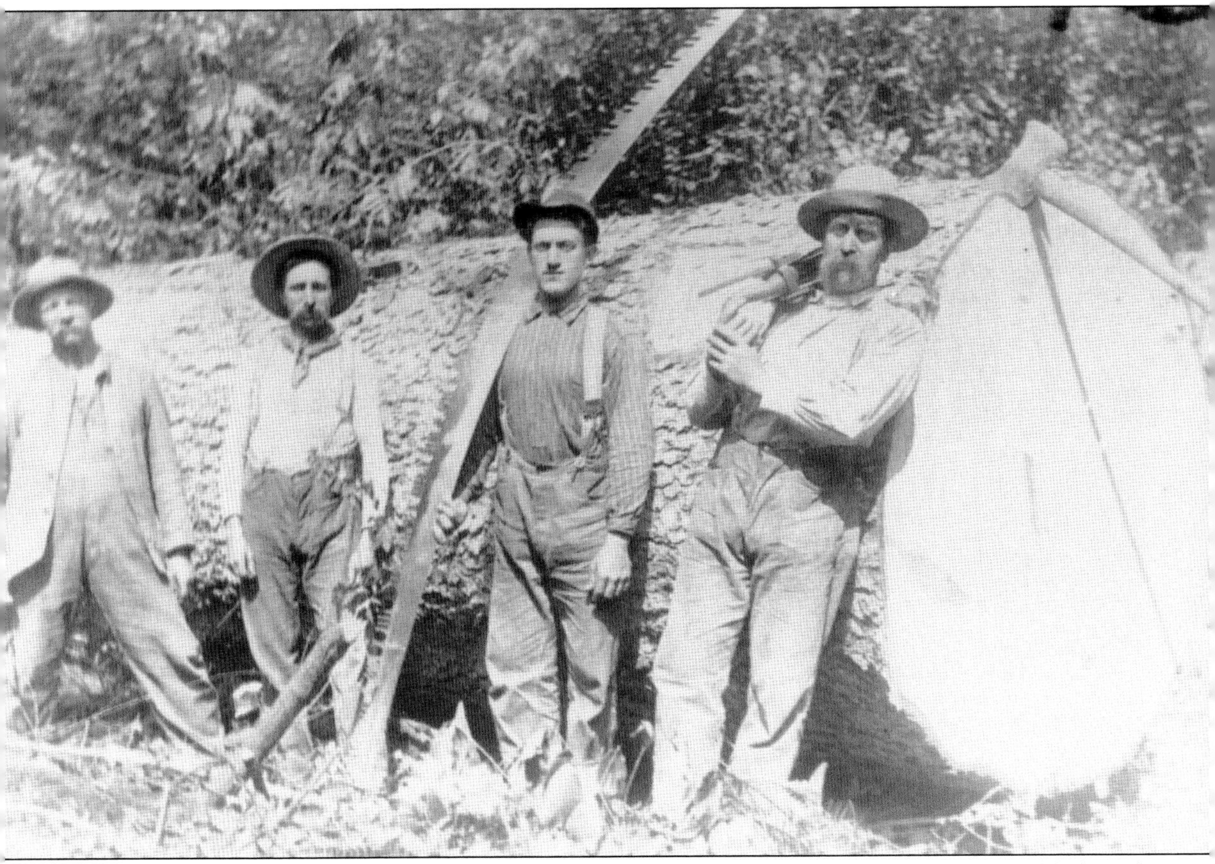

Lumberjacks used two types of axe in the 1800s: a single-bitted (or blade) axe and a double-bitted axe. Some loggers believed the double-bitted axe, as seen above, was a dangerous choice, preferring to use two single-bitted axes instead. One axe was used to get through the bark, and the second to fell the tree. Up until the 1880s, lumberjacks felled trees with axes exclusively. The custom of using the crosscut, or "misery whip," saw began among loggers in Pennsylvania and spread from there. The nickname for this tool comes from the difficulty and frustration of using a saw that could not be kept sharp enough. Misery whips came in a variety of sizes, ranging from the one-man saw (from three to six feet) to the two-man saw (from four to 16 feet). The flexible and relatively light crosscut saws, like the one seen here, were also called felling saws. (Courtesy of the Roger Fox collection.)

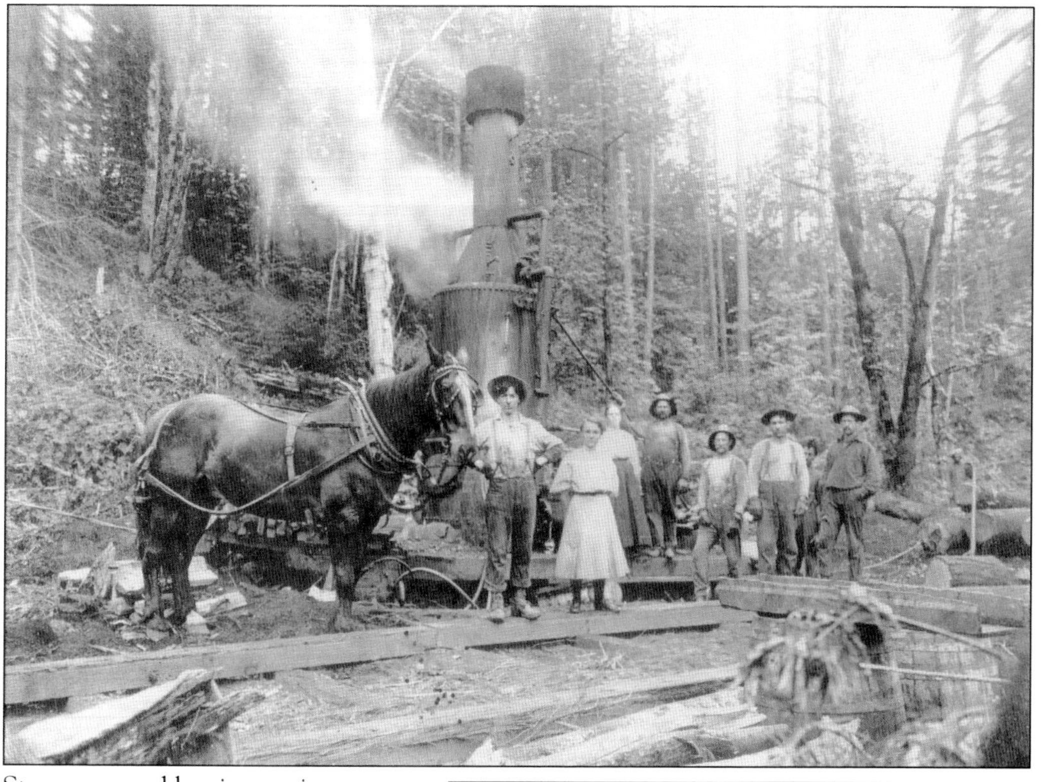

Steam-powered logging engines, like the one seen here with Red Blackmore and his wife, Ora, revolutionized the logging industry. The steam-powered winch, commonly referred to as a "steam donkey," used ropes or cables to pull logs for distances of up to a mile. This increased the amount of forest that could be logged while decreasing the amount of man or animal power required. (Courtesy of the Roger Fox collection.)

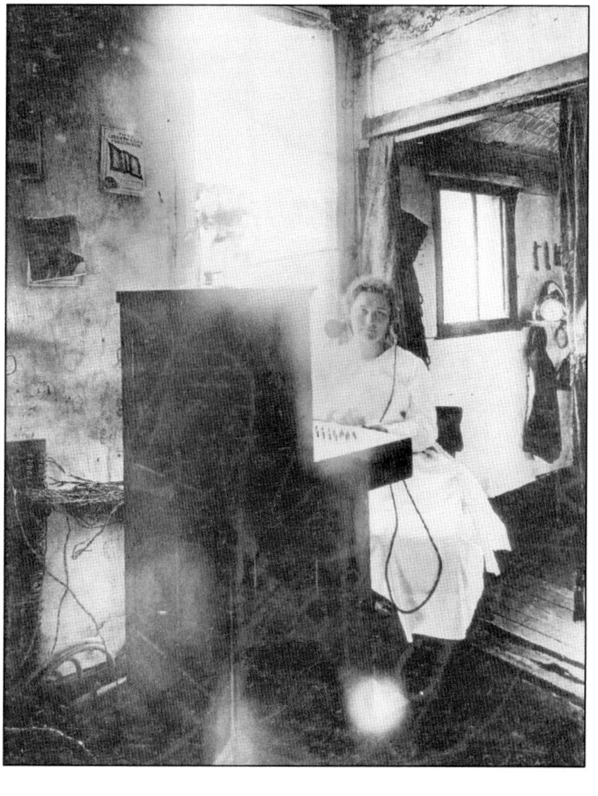

In 1886, the Pacific Postal Telegraph Company established a line through Mount Vernon to Whatcom. The first operator was Thomas Payne, and the first telegraph office in Mount Vernon was located in Hartson's printing office. In 1894, the first telephone service arrived in Mount Vernon thanks to the Sunset Telephone & Telegraph Company. Seen here is the unidentified first operator. (Courtesy of the Roger Fox collection.)

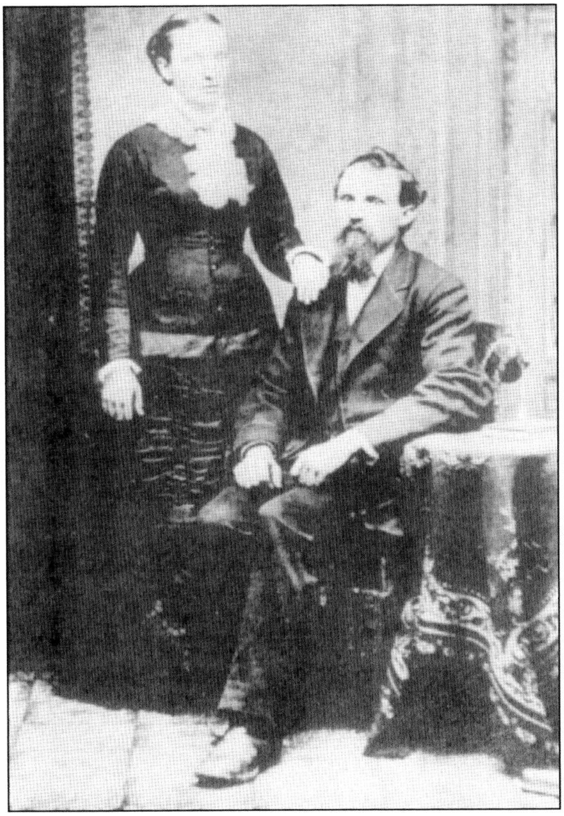

Rev. Brisbon Napolean LaJeunesse Davis (pictured with wife Doliska) was born in East Tennessee on November 20, 1849, and became the first permanent settler on the upper Skagit River, above the logjam, in 1873. He founded First Baptist Church, Mount Vernon's first church, on April 27, 1884. He also introduced the first Holstein cattle into Skagit County, importing purebred, registered Holsteins in 1885. (Courtesy of Mount Vernon First Baptist Church.)

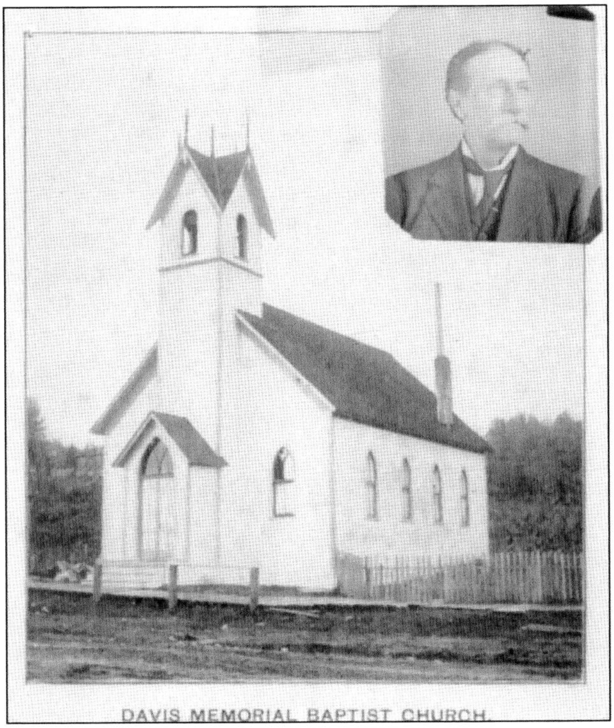

The original congregation of First Baptist Church had only four members, but by 1885, the congregation had grown and obtained two lots at the corner of First and Division Streets. The new church, shown here, was dedicated on November 17, 1889. Also pictured is Rev. Solomon A. Abbott, who revitalized the congregation after the untimely death of B.N.L. Davis in 1891. (Courtesy of Mount Vernon First Baptist Church.)

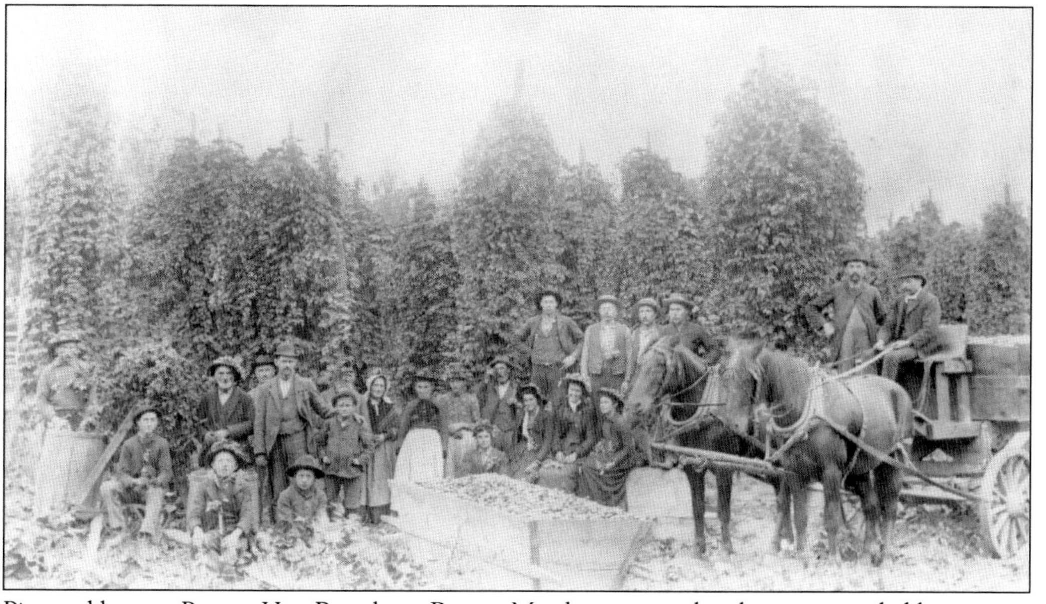

Pictured here at Powers Hop Ranch on Beaver Marsh are several pickers surrounded by massive hop plants. The hops were harvested by locals as well as by Indians from British Columbia. One year, Rev. B.N.L. Davis netted approximately $40,000 from a hops crop and used the money to bring the first purebred cows and registered Holstein bulls to Skagit Valley in 1885. (Courtesy of the Skagit County Historical Society.)

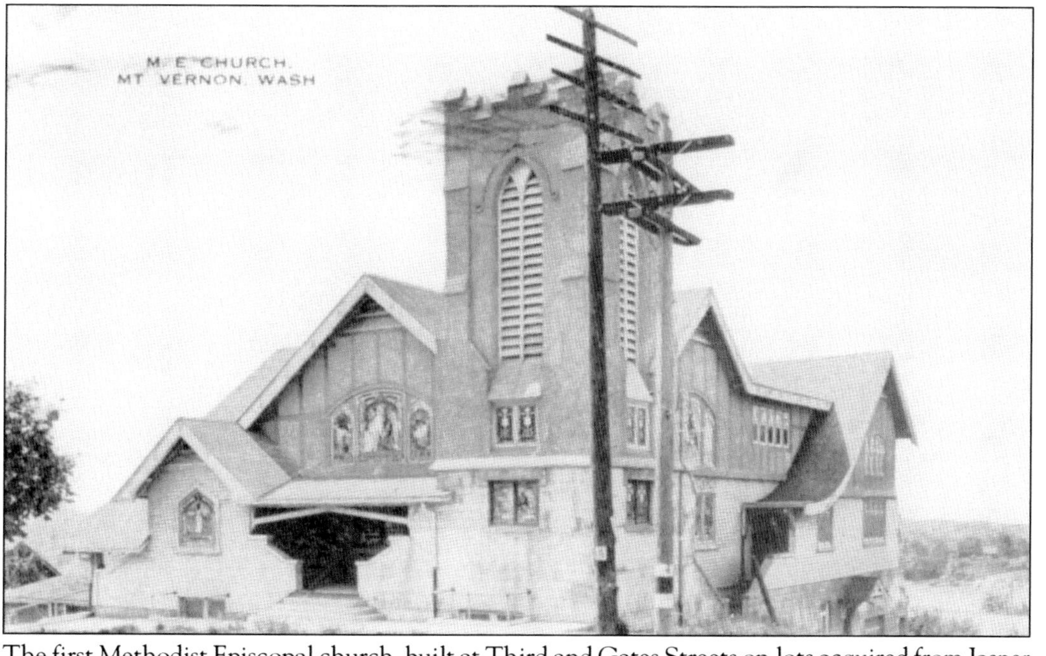

The first Methodist Episcopal church, built at Third and Gates Streets on lots acquired from Jasper Gates, was dedicated in 1890 and served the congregation until 1910. The downtown building was sold in 1911, and construction of a new structure began at the corner of Fourth and Fulton Streets. The new church, seen here, was completed and dedicated in 1915. The parishioners worshiped here until they relocated in 1960. (Courtesy of John Kamb Jr..)

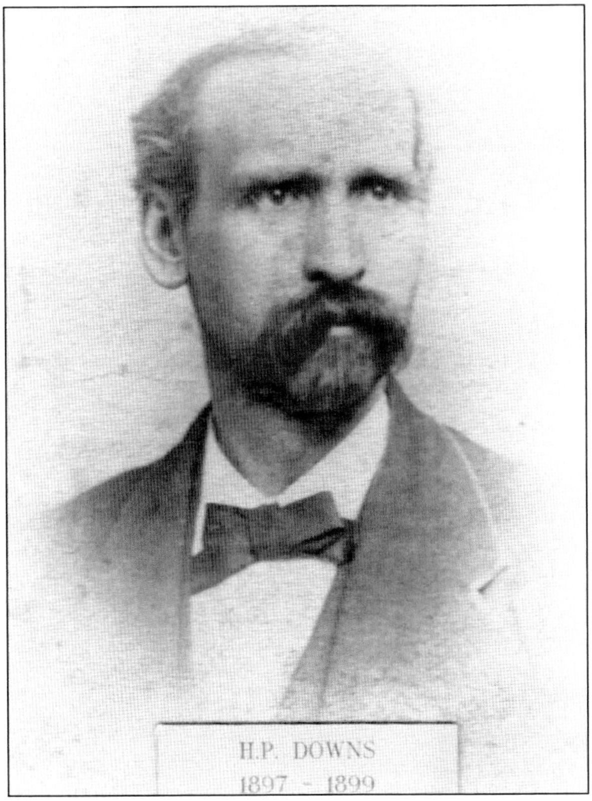

Dr. Horace P. Downs, a physician from Boston, came to the Skagit region in 1878. He was chosen many times for elected offices, including county commissioner, county assessor, deputy assessor, and mayor (twice, 1897–1899). Downs was the Skagit County auditor who certified the June 27, 1890, election in which incorporation was approved and Charles D. Kimball was elected as the city's first mayor. (Courtesy of the Mount Vernon Mayor's Office.)

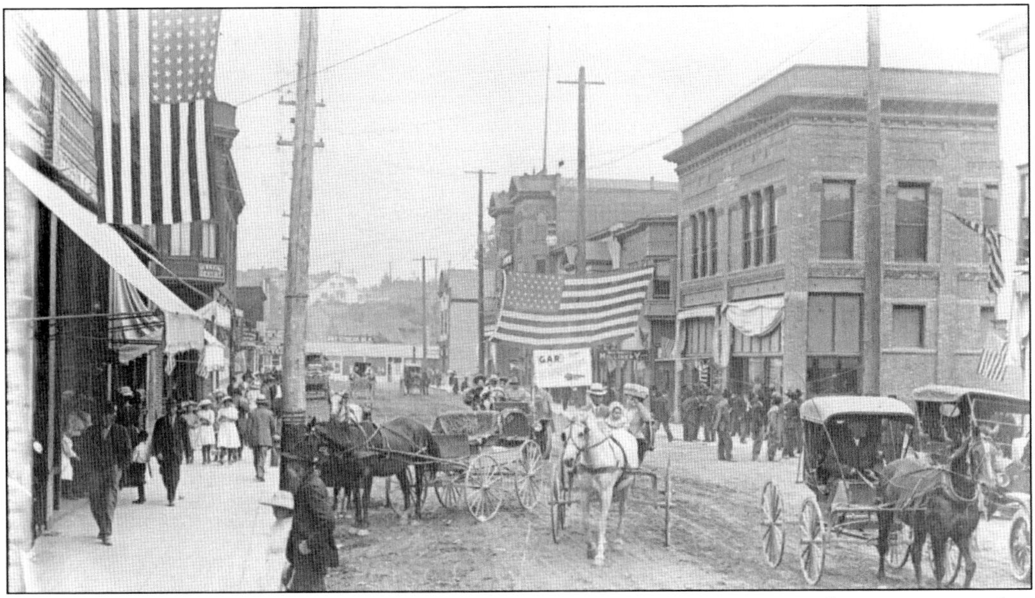

In the summer of 1892, Downs constructed the Opera House, an ornate three-story wooden building that had a 23-by-50-foot stage, a 30-foot-high auditorium, and seating for 850 people. Inexplicably, photographs of the Opera House are rare, but it can be seen in the center of this early-1900s street scene, the top of the building above the large American flag. (Courtesy of John Kamb Jr..)

Three

Overcoming Adversity 1891–1899

The final decade of the 19th century brought phenomenal highs and decimating lows to the fledgling city of Mount Vernon. After its meteoric rise, both fires and floods brought everything back to earth. On July 13, 1891, a devastating fire destroyed most of the riverside commercial district, and many of the businesses subsequently relocated to First Street, away from the river. Major floods in 1892 and 1894 motivated the city to initiate construction of a massive dike system in the summer of 1894. And if Mother Nature was not a big enough problem, the country was in the grips of an economic depression brought on by the Panic of 1893.

However, all that destruction could not stop the progress that was still enveloping the area. On August 12, 1891, the Great Northern Railroad reached Mount Vernon, giving the city another vital connection to the outside world. That same summer, the Mount Vernon Electric Company (the first electric plant in all of Skagit County) was established. Lincoln School, the town's first brick school building, was constructed in 1891. Its location on "the Hill," east of town, was a reflection of a shift in population as more homes were being built in that area. The Louvre Hotel, the city's first brick hotel, was erected during this period, and in the summer of 1892, the Opera House made Mount Vernon a cultural center in the county. In 1893, the city's first bridge across the Skagit, a wooden-truss structure with a draw-span to allow vessels to pass, was opened.

In November 1892, there was another county seat election. Although Mount Vernon came in second place to Anacortes (867 versus 873 votes), changing the county seat required a three-fifths majority vote, so the distinction remained with Mount Vernon. In 1893, the county records, which had remained in La Conner since the first election, were moved to the newly constructed courthouse in downtown Mount Vernon.

Fortunately, the abundant natural resources of the area allowed Mount Vernon to weather the country's financial crisis well. The town closed out the 19th century on a steady course of growth.

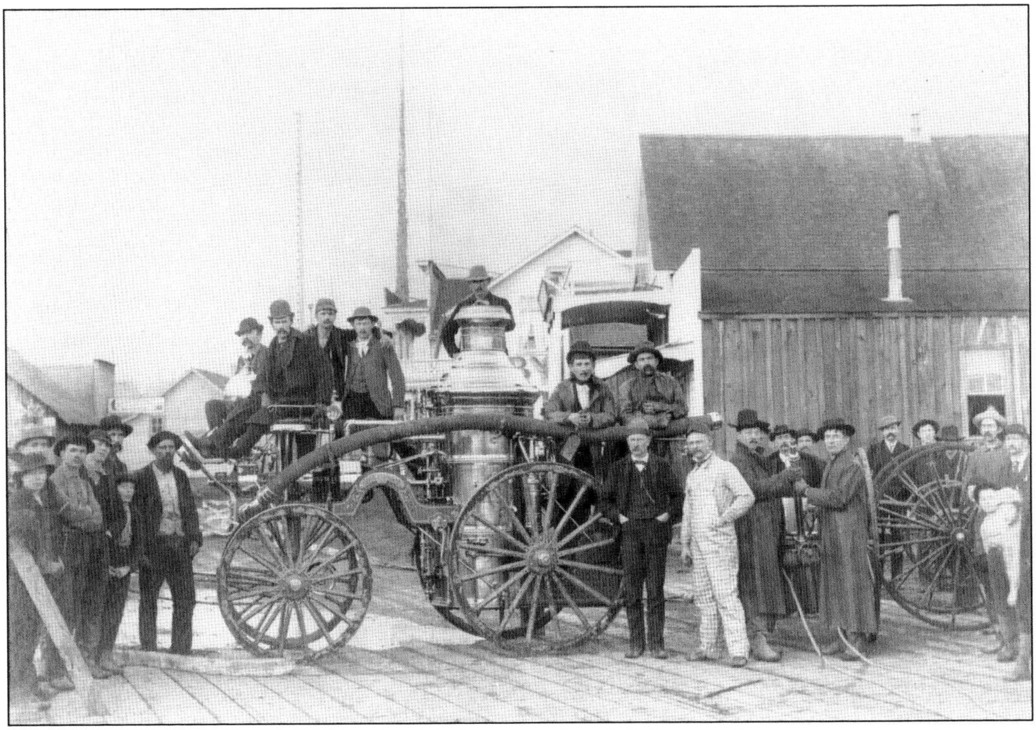

On July 13, 1891, a fire started in the Washington Hotel. The wooden structure was destroyed, along with 15 other business buildings and two residences, before the fire could be brought under control. The volunteer-operated fire equipment, seen above, broke down during the blaze. It was the crew of the steamer *Baily* who saved the town from further destruction. The first fire department was formed in 1891, consisting of, from left to right, (first row) M. Haines, Frank Noyes, Duke Elkins (assistant chief), Warren Haskell (chief), M. Hungerford, Clyde Anable, Ira Paterson, and John Anabel; (second row) Charles Kimble, Billy Kline, Frank Reeves, Bernie Jarvis, Ed Kimble, Fred Pickering, George Faxon, and Napoleon Fortin; (third row) Frank Birchie, Billy Pucket, Joe Mason, Shell Elkins, Joe Fredlund, Billy Anable, and Frank Beal. (Both, courtesy of the Roger Fox collection.)

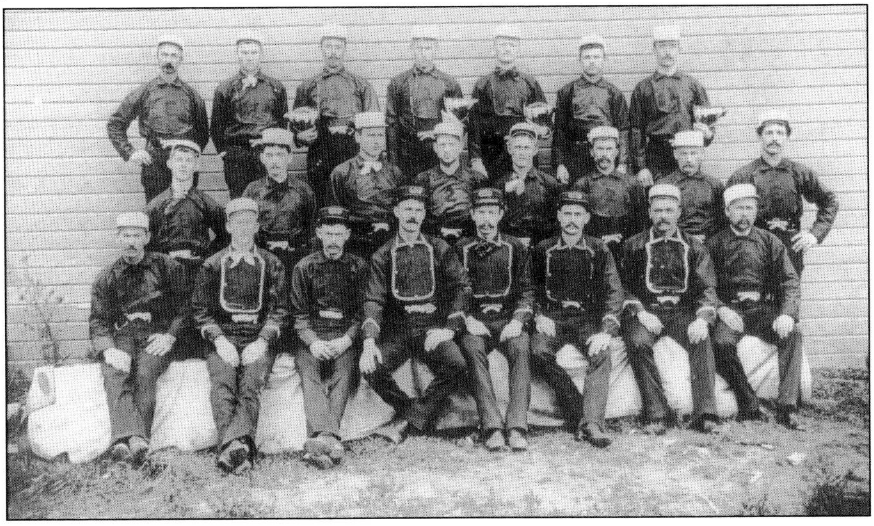

Seen here is the Priest Millinery, located downtown, which was run by the widowed Ida Priest. Priest had come west from Wisconsin and was a prominent citizen in the city. She was the only woman to help present the petition for incorporation to the Skagit County commissioners and district court judge Winn in 1890. (Courtesy of the Roger Fox collection.)

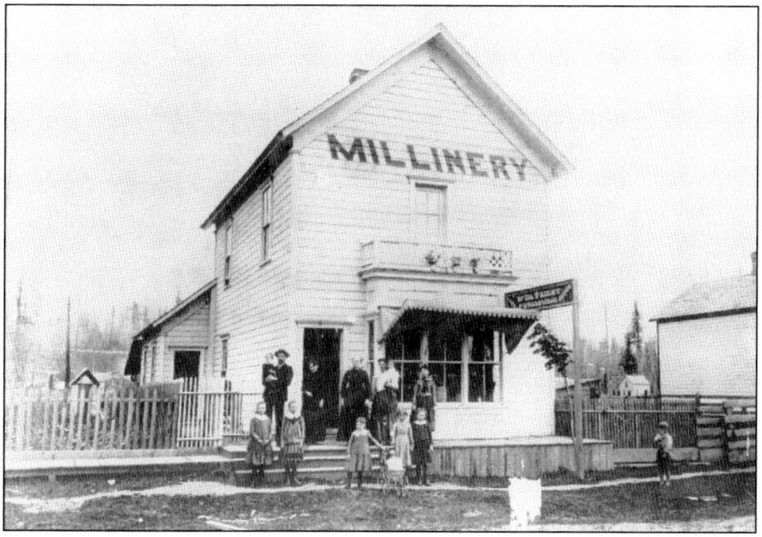

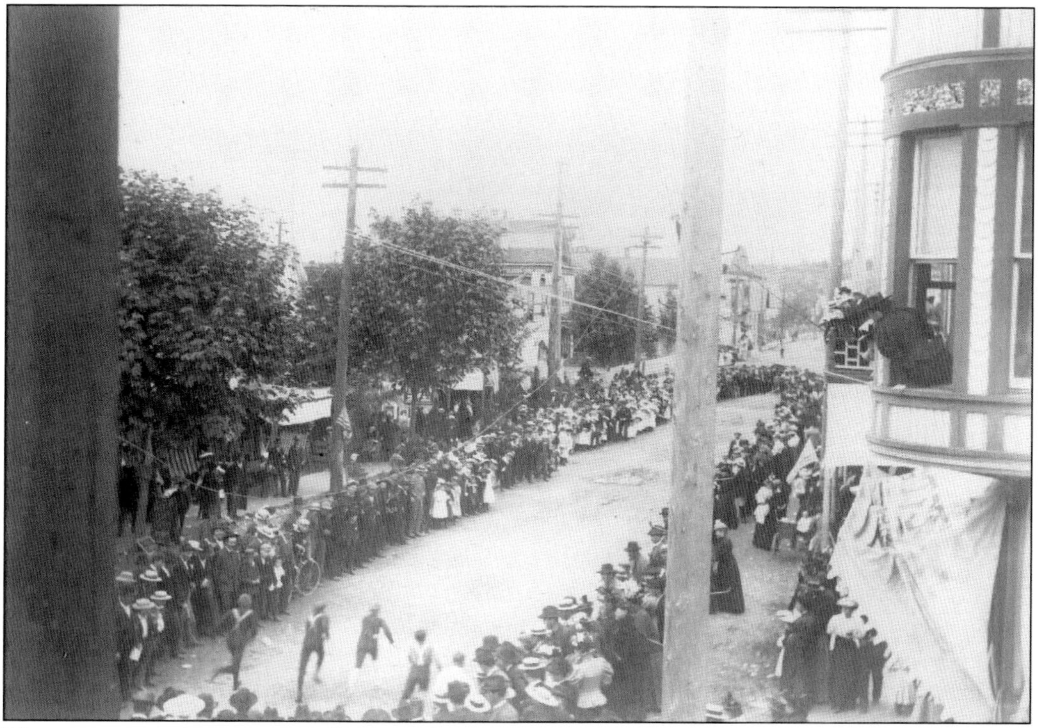

These men are taking part in a foot race down First Street around 1890. The race was probably held in conjunction with a local celebration, judging from the large crowd gathered. The photograph was taken from a second-story or rooftop location, looking north. The lady on the right is watching from an upstairs window of the Miller Hotel. (Courtesy of John Kamb Jr..)

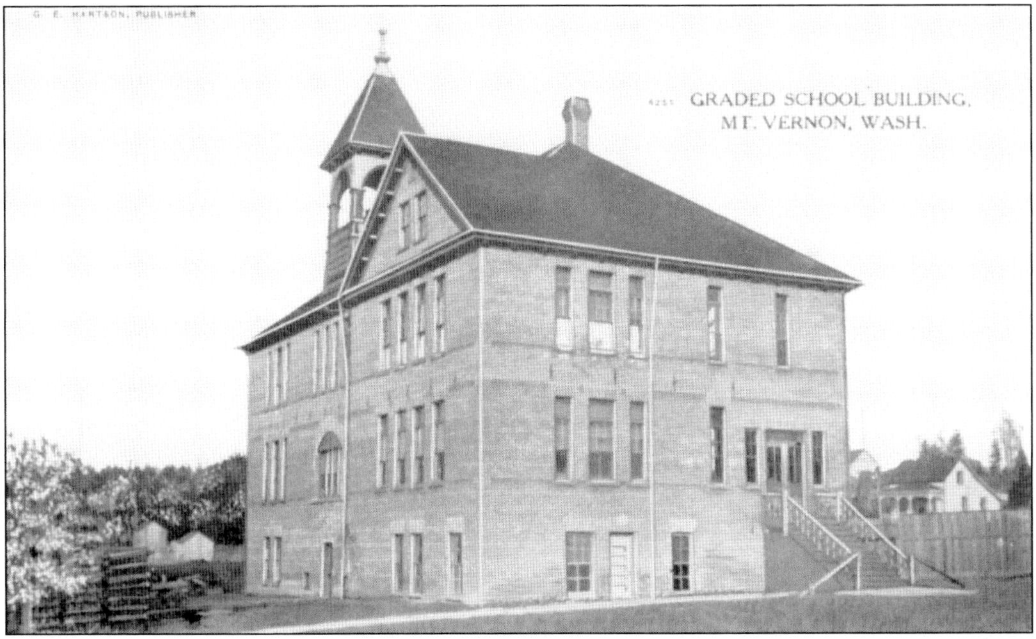

Lincoln School, constructed in 1891, was the first brick school building in Mount Vernon and was one of only two in Skagit County (the other being Columbian School in Anacortes). The high board fence, seen on the right of the image, separated the girls' playground from the boys'. Up until 1895, the school system in Mount Vernon only went through the ninth grade. The picture below, taken in the early 1900s, shows a group of unidentified teachers in one of Lincoln's classrooms. The school was used exclusively as an elementary school from 1908 until it was torn down in 1938. A new Lincoln Elementary School was built that same year in the same location and is still used today. (Above, courtesy of John Kamb Jr.; below, courtesy of the Roger Fox collection.)

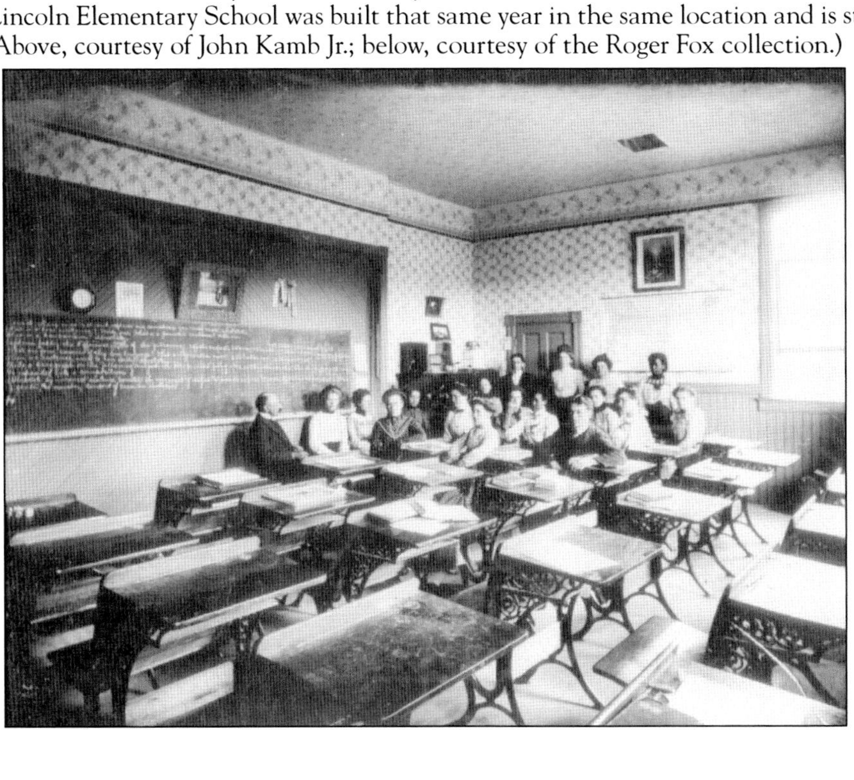

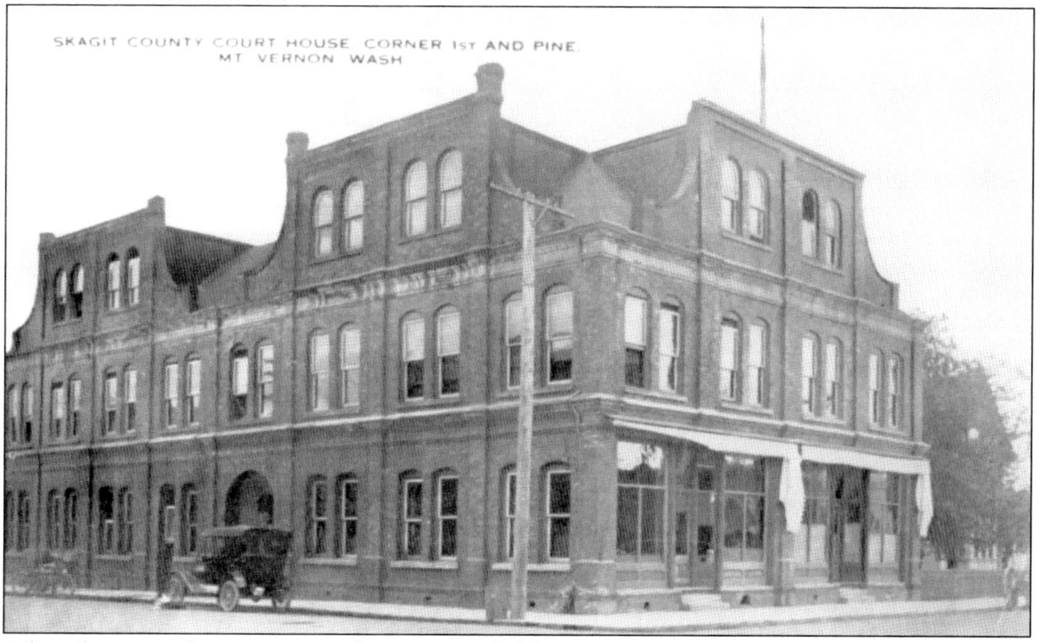

After the second county seat election was held in November 1892, the city lost little time beginning construction of a county courthouse. At the corner of First and Pine Streets, on land that had belonged to Capt. D.F. Decatur, a 50-by-114-foot building was constructed for $13,585. It would serve as the courthouse, jail, and offices of county government until 1922. (Courtesy of the Roger Fox collection.)

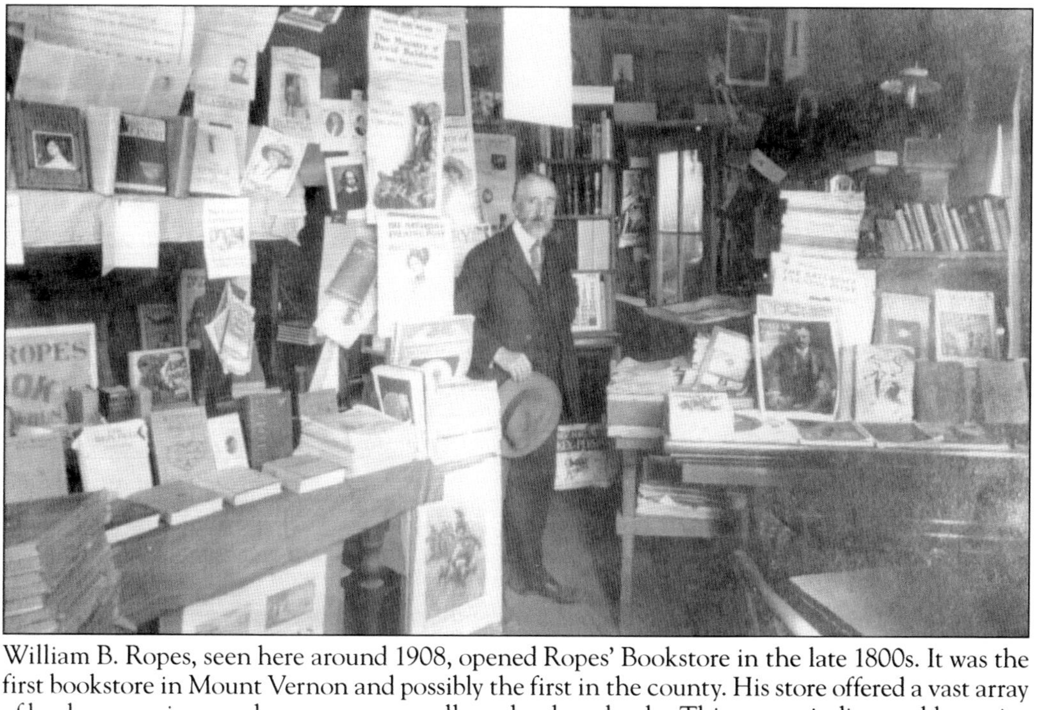

William B. Ropes, seen here around 1908, opened Ropes' Bookstore in the late 1800s. It was the first bookstore in Mount Vernon and possibly the first in the county. His store offered a vast array of books, magazines, and newspapers as well as school textbooks. This was an indispensable service for a city that did not have a library until 1916. (Courtesy of the Roger Fox collection.)

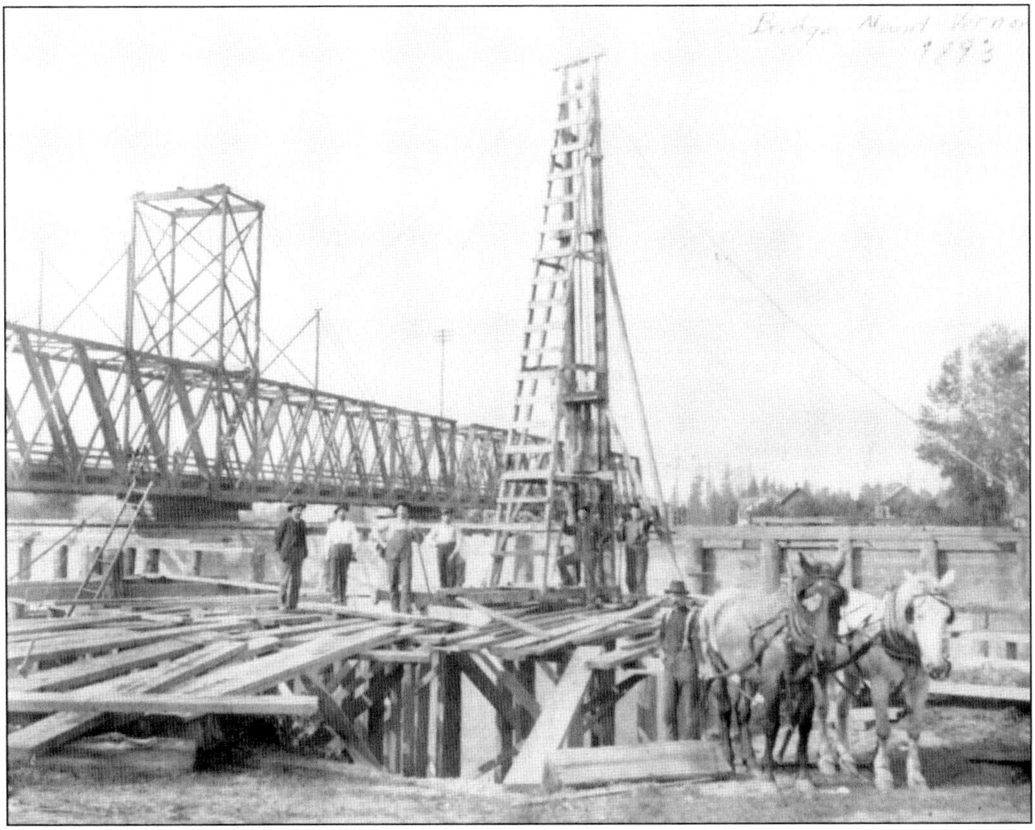

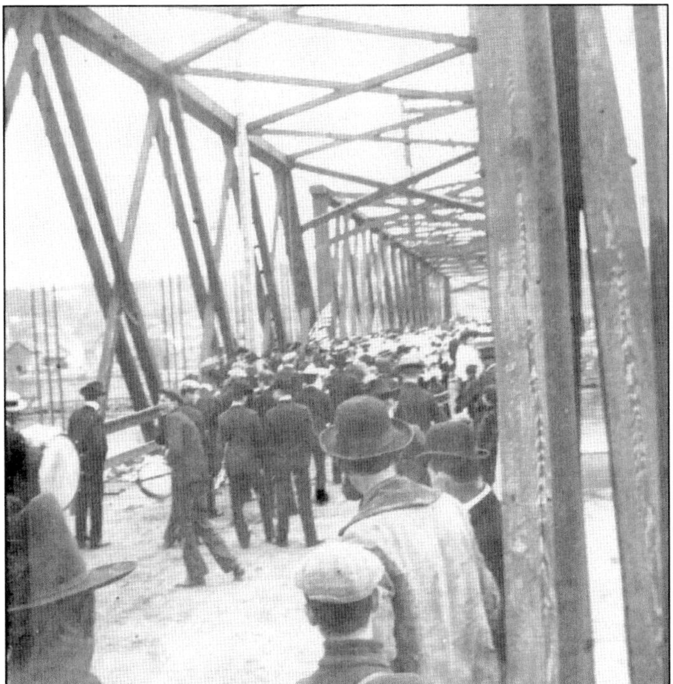

Above, unidentified workers construct the first bridge to cross the Skagit River at Mount Vernon. The bridge, pictured at left on Decoration Day, opened in 1893. The wooden-truss structure had a draw-span that had to be opened by hand. At each end of the bridge was a sign reading "$25 fine for riding or driving over this bridge faster than a walk." During the flood of 1906, the bridge was damaged beyond repair. It was replaced in 1908 with a bridge designed to accommodate automobiles. The new structure had a swing-span powered by a gasoline motor. (Above, courtesy of the Mount Vernon Mayor's Office; left, courtesy of John Kamb Jr..)

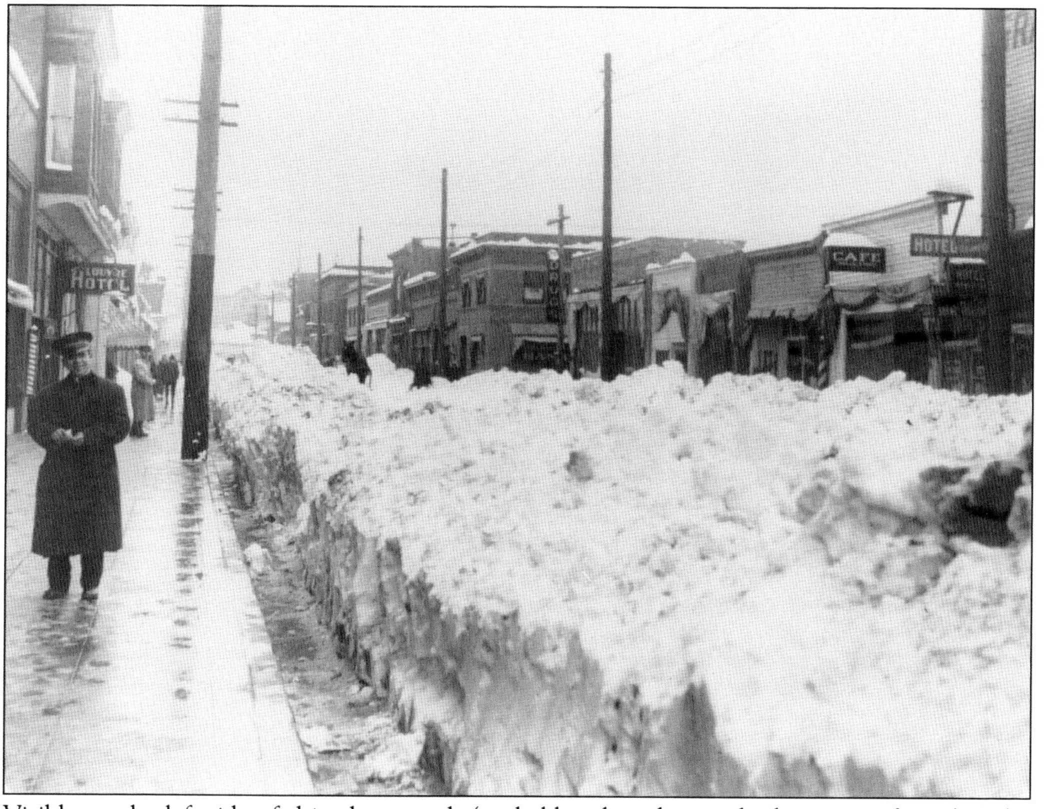

Visible on the left side of this photograph (probably taken during the big snow of 1916) is the Hotel Louvre, the town's first brick hotel, which was built in the 1890s. It was "the most popular place to stop at while in Mount Vernon" until 1910, when the much grander Windsor Hotel was built, just south and across the street. In 1911, proprietor William Anable spent $6,000 to update the Louvre, fitting 26 apartments, parlors, offices, and bathrooms on the second floor and refurbishing the lower floor with dining parlor, pool room, hotel office, and cigar and candy hall. Pictured below are Bob Davis (left) and Fred Rhodes (right) in the Louvre Pool Hall around 1920. The hotel operated until the 1960s. (Above, courtesy of the Mount Vernon Mayor's Office; below, courtesy of the Roger Fox collection.)

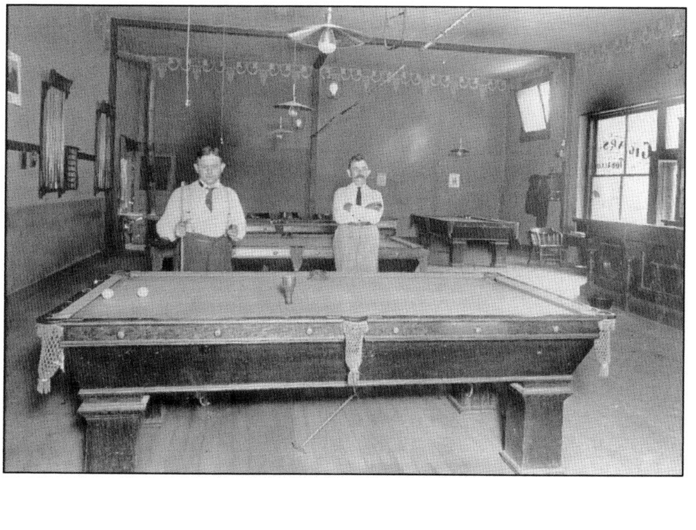

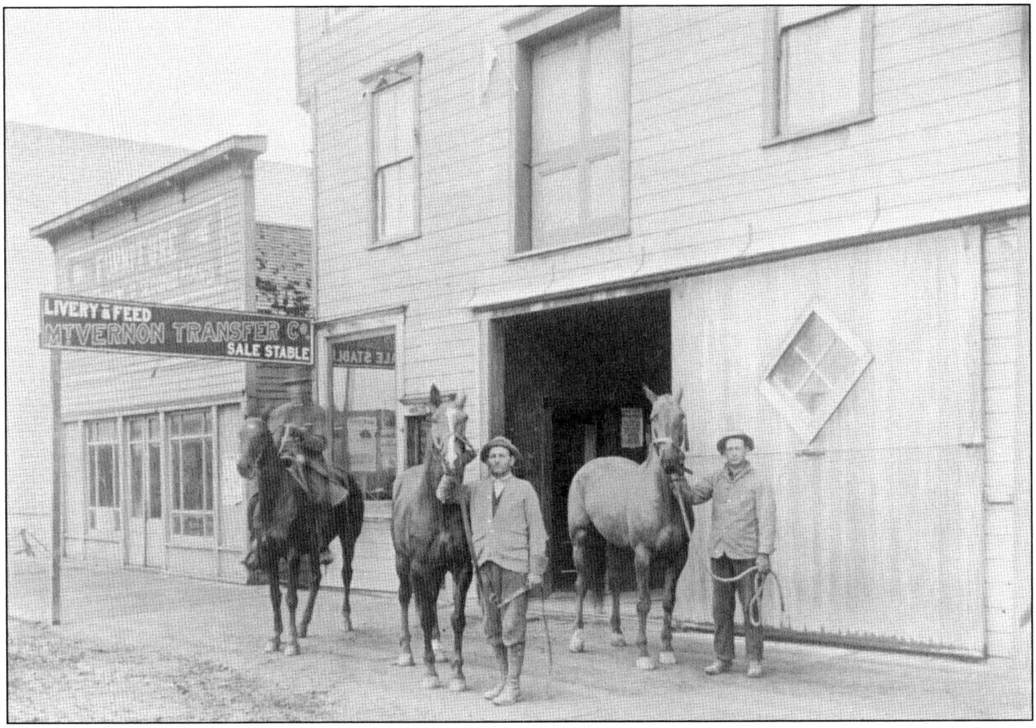

Livery stables, like the one seen here, were places where horses, teams, and wagons could be hired and where privately owned horses could be boarded for a short time. They were often found attached to a hotel or boardinghouse. The livery stable was a necessary institution of every American town, providing vital transportation services and a source of hay, grain, coal, and wood. (Courtesy of the Mount Vernon Mayor's Office.)

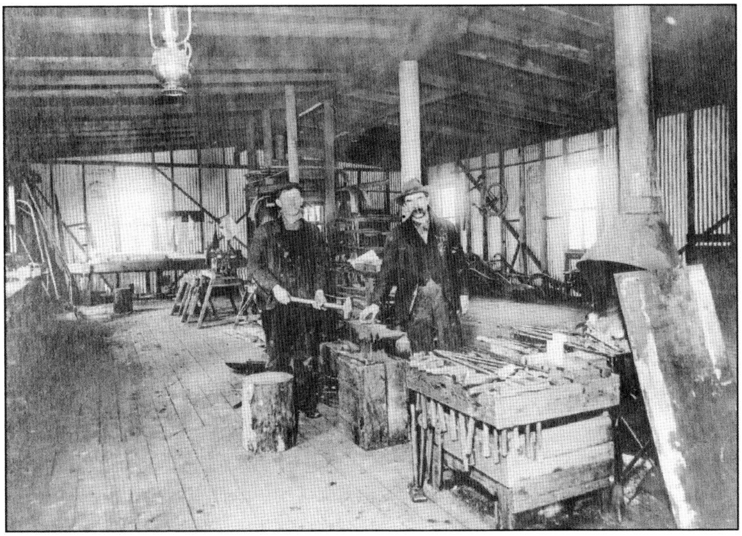

Pictured in the late 1800s, this is the blacksmith shop at the Mount Vernon waterfront, near the east end of the bridge. On the left is L. Fredman (stockkeeper), and on the right is Oscar Higman. The importance of blacksmiths in the days when horses were a major mode of transportation cannot be overstated. (Courtesy of the Roger Fox collection.)

Four

FARMS, FOUNDATIONS, AND FIRE 1900–1911

According to the 1900 federal census, Mount Vernon had a total of 1,120 citizens. Logging had gradually moved away from the city, and the area was by now strongly agricultural. While the initial agricultural lands were used mostly for the production of oats and hay, now dairy farms constituted the majority of farming. In 1905, it was estimated that 3,000 cows could be found within a six-mile radius of city. The next year, the Mount Vernon Cream Company was built on the east bank of the Skagit River, and a few months later the Pacific Coast Condensed Milk Company (later the Carnation Company) was completed. It is clear to see how Mount Vernon earned the nickname "Milk City."

The city continued to expand a thriving infrastructure. The Skagit County Fairgrounds were established in 1901. In 1905, the first city officers were elected, and in 1907 the first city council meeting was held. Washington School, a two-story wood-frame building housing grades one through eight, was constructed in 1905. It was quickly followed by Roosevelt in 1908, which served as a four-year high school. The Commercial Club (later to become the chamber of commerce) was organized in 1903. In 1904, rural mail delivery was established. A sewer system was built in 1909, and the first street paving was laid in 1911.

Fires continued to plague the city. On September 28, 1900, another major fire in the original business district of the city destroyed most of the remaining pioneer-era wooden commercial buildings. Fortunately, the citizens were turning to sturdier, fire-resistant brick construction. The Union Block Building, constructed in 1906 across Pine Street from the county courthouse to serve as legal offices, still stands today. In 1909, the luxurious three-story Windsor Hotel was built. This building also housed the Mount Vernon State Bank, which had been established in 1906.

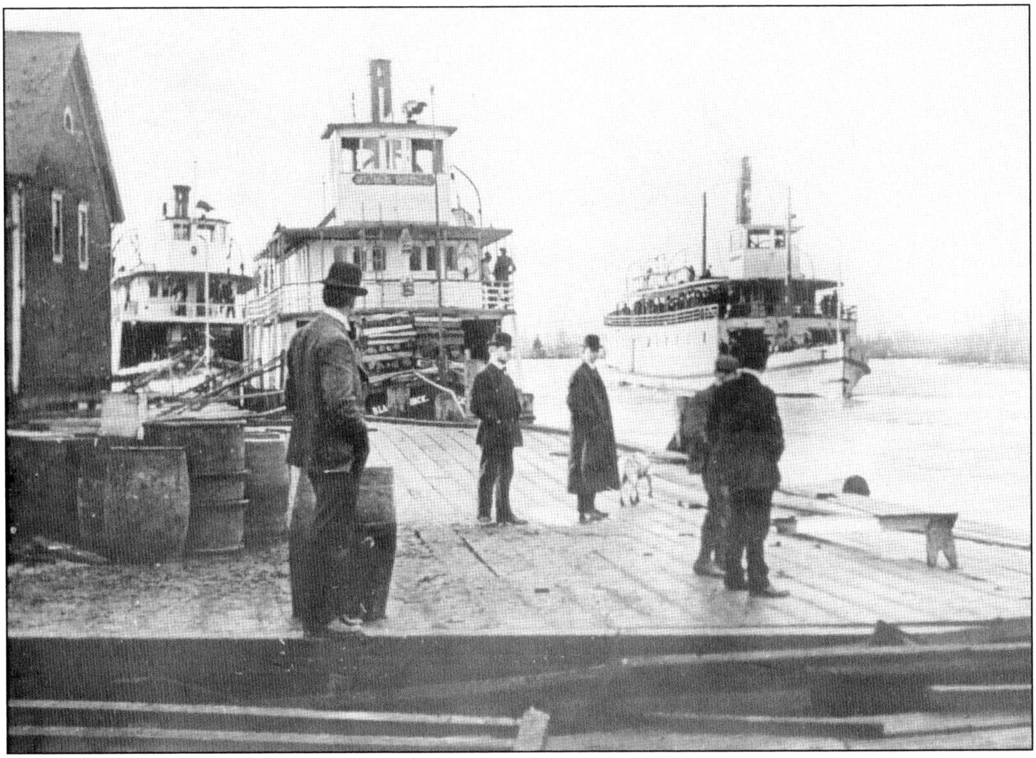

Early merchants used the Skagit River to trade with other cities and towns. Floating docks, as pictured above, were built along the riverbanks in Mount Vernon and provided a place where ships like the *Black Prince* (center) could tie up to load and unload their cargo. Third from the left is Nelson J. Moldstad, who opened a clothing and dry goods store at the corner of First and Gates Streets with Louis Foss after their 1891 arrival in Mount Vernon. The store can be seen in the street scene below in the center left. Moldstad became the president of the Skagit County Fair Association and was elected to the state legislature as a Republican in 1902. (Above, courtesy of the Roger Fox collection; below, courtesy of John Kamb Jr..)

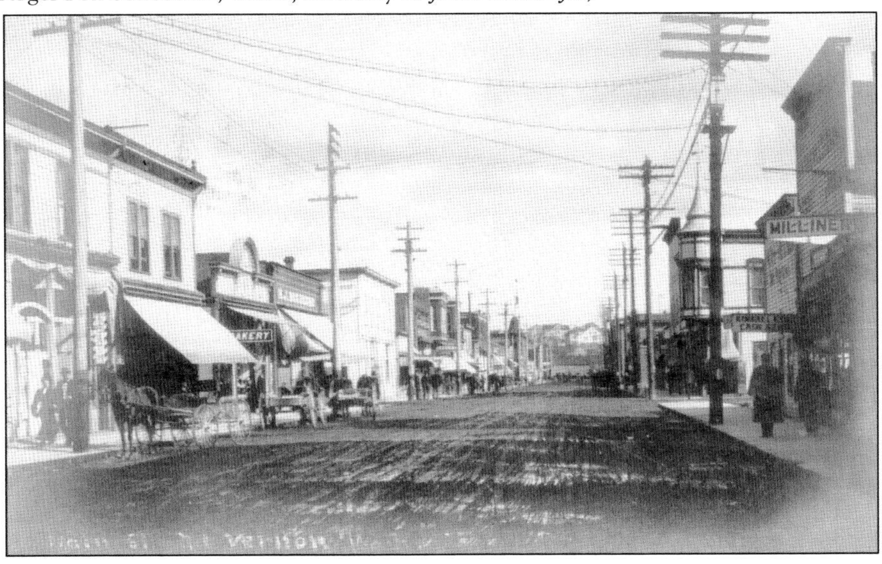

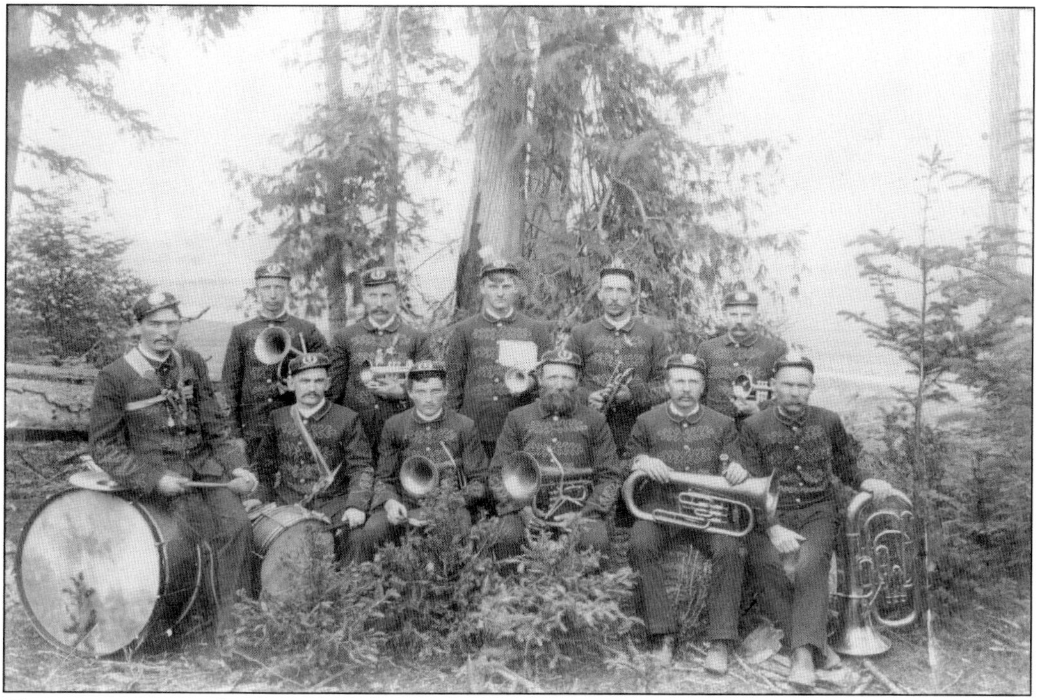

This group, pictured around 1900, is referred to as "Mount Vernon's first band." The musicians are, from left to right, (first row) Charles Gilkey, Lew J. Bailey, Horace Neeley, Dennis Storrs, John Anable, and George E. Hartson; (second row) Arthur Storrs, John Garl, Albert Neeley, Newt Gates, and R.L. Martin. (Courtesy of the Roger Fox collection.)

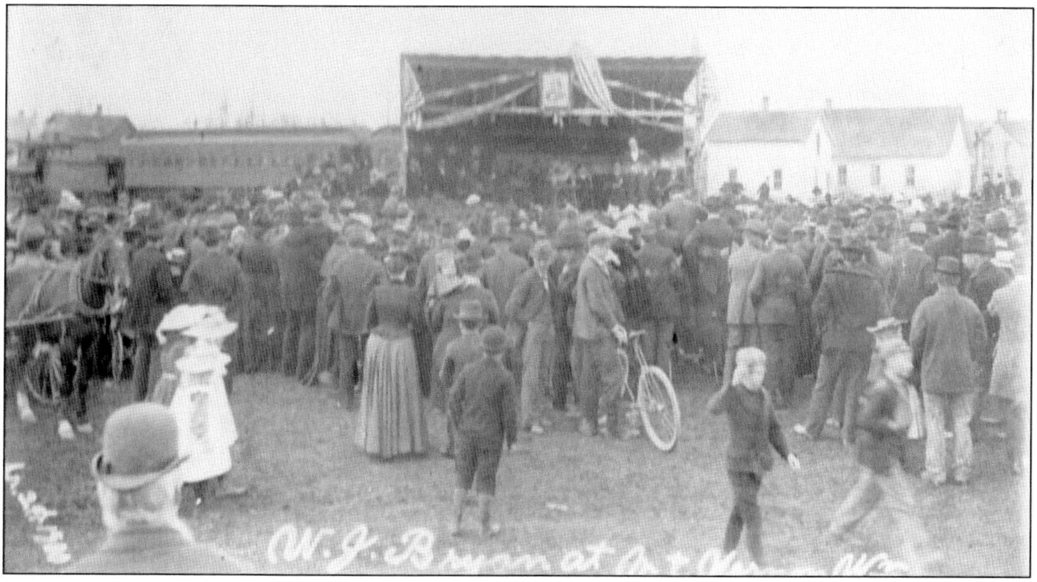

William Jennings Bryan (March 19, 1860–July 26, 1925) was a three-time Democratic nominee for president of the United States (in 1896, 1900, and 1908). With his deep, commanding voice and extensive travels, he was one of the best-known orators and lecturers of the period. This photograph shows him campaigning in a fledgling Mount Vernon in 1900. He would lose that election to William McKinley. (Courtesy of John Kamb Jr..)

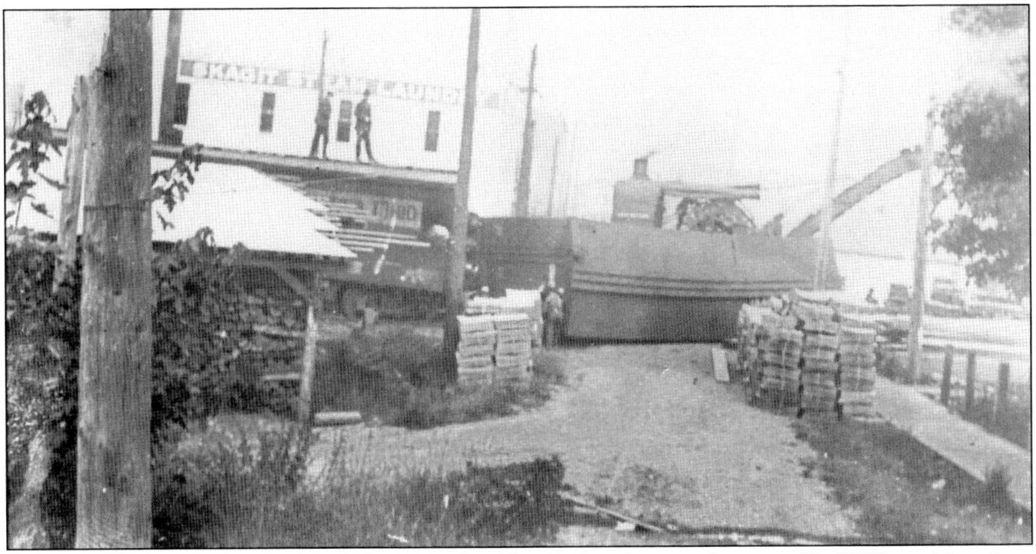

In 1903, Oscar Sundstrom (born in West Gothland, Sweden, on January 17, 1872) and brother-in-law Adolph Anderson started the Skagit Steam Laundry in Mount Vernon. The laundry, seen above, was located at the bottom of the viaduct on Washington Street. It is visible in this westward view of Washington Street, along with the steam crane on Third Street. Anderson's half interest in the laundry was later purchased by Charles Clinton Johnson, who was a successful Skagit County farmer before he turned to business. The image below shows large, steam-powered mangles and unidentified staff inside the steam laundry around 1920. (Above, courtesy of John Kamb Jr.; below, courtesy of the Roger Fox collection.)

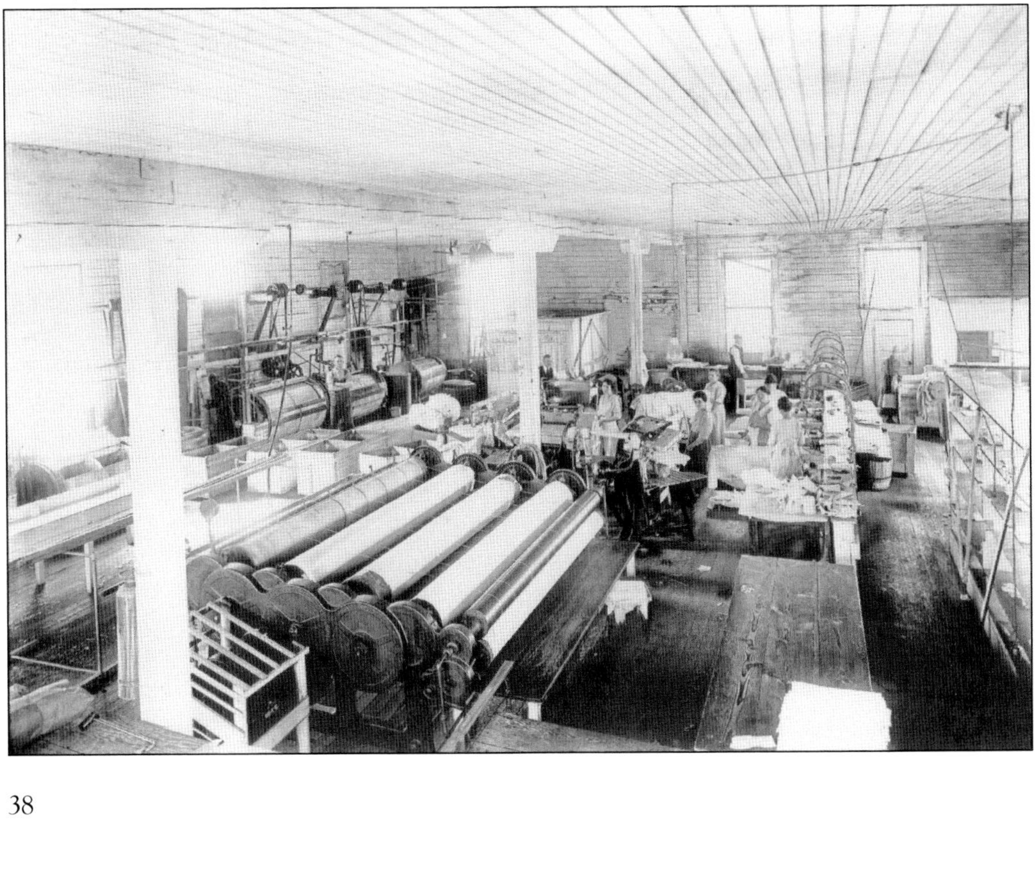

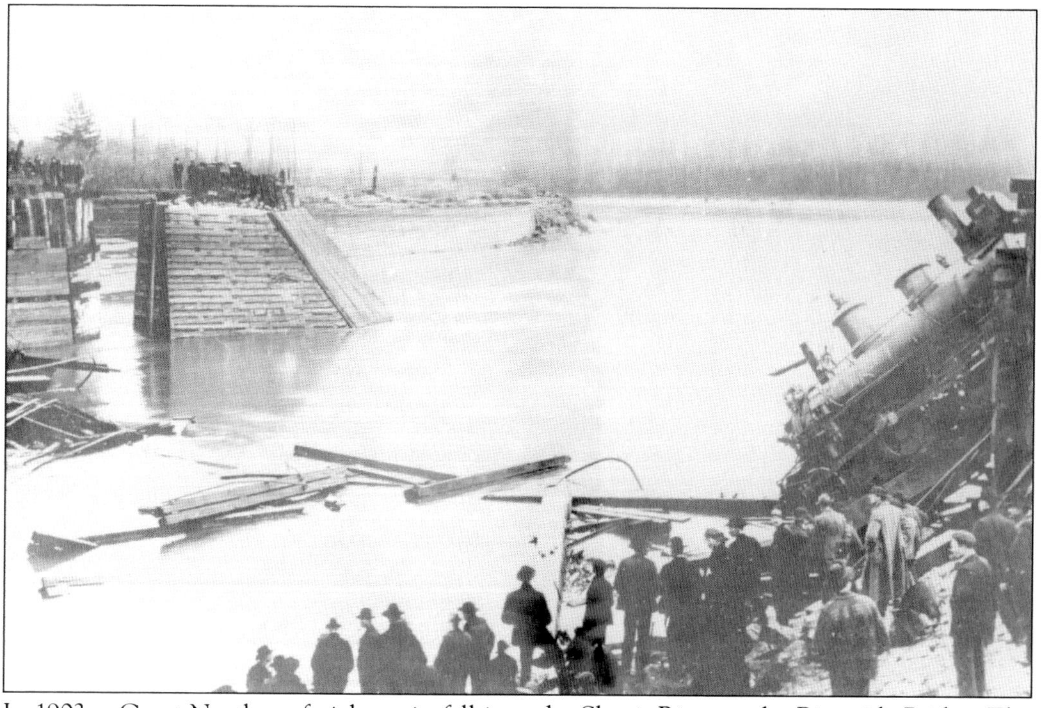

In 1903, a Great Northern freight train fell into the Skagit River at the Riverside Bridge. The locomotive had just reached the south bank when the bridge collapsed and the cars pulled the engine back into the river; one car remained on the north side of the bridge. An earlier derailment of a car of logs was responsible for weakening the bridge. (Courtesy of Ed Marlow.)

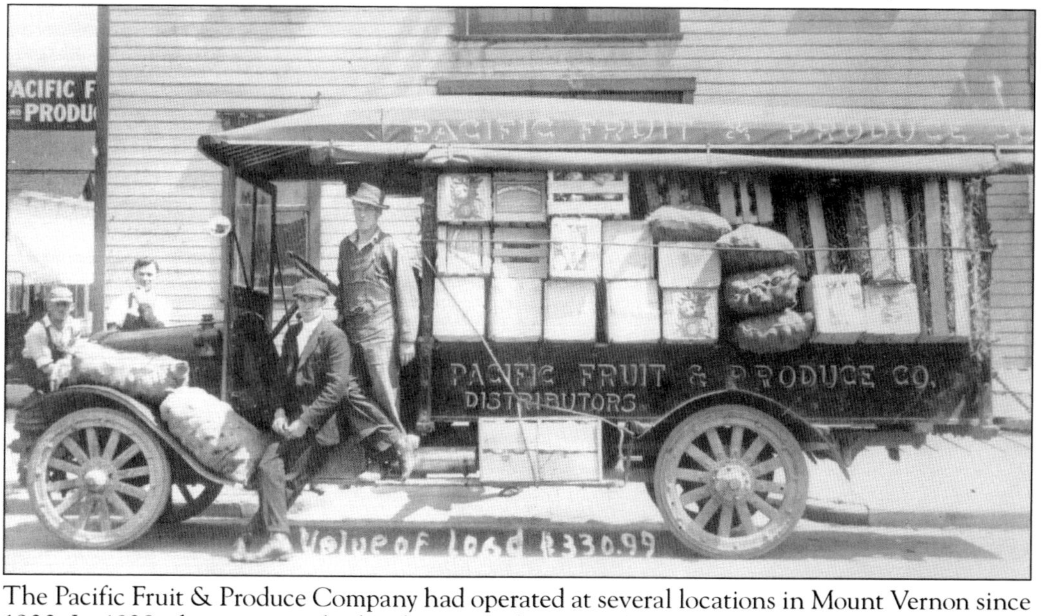

The Pacific Fruit & Produce Company had operated at several locations in Mount Vernon since 1900. In 1928, the company built a large warehouse and distribution center on Third Street. The company provided a continual source of fruit and produce for grocery stores, restaurants, and institutions over a wide region. It also purchased produce from local growers to ship to other distribution points. (Courtesy of the Roger Fox collection.)

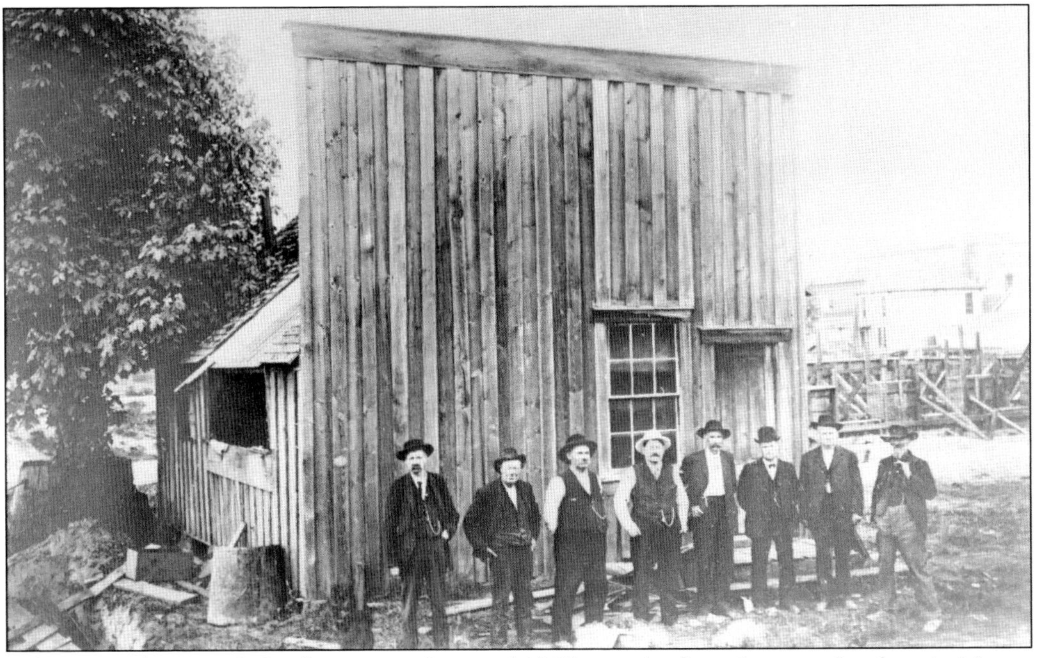

Robinson's Photograph Studio, pictured around 1905, was the first photography shop in Mount Vernon. Pictured are, from left to right, A.M. Elkins (blacksmith), W.F. Robinson (photographer), George E. Hartson (newspaperman), George Moran (hotel keeper), Charles Harmon (politician and farmer), George Morlan (storekeeper), Neil Mulvey, and William Woodward. (Courtesy of the Roger Fox collection.)

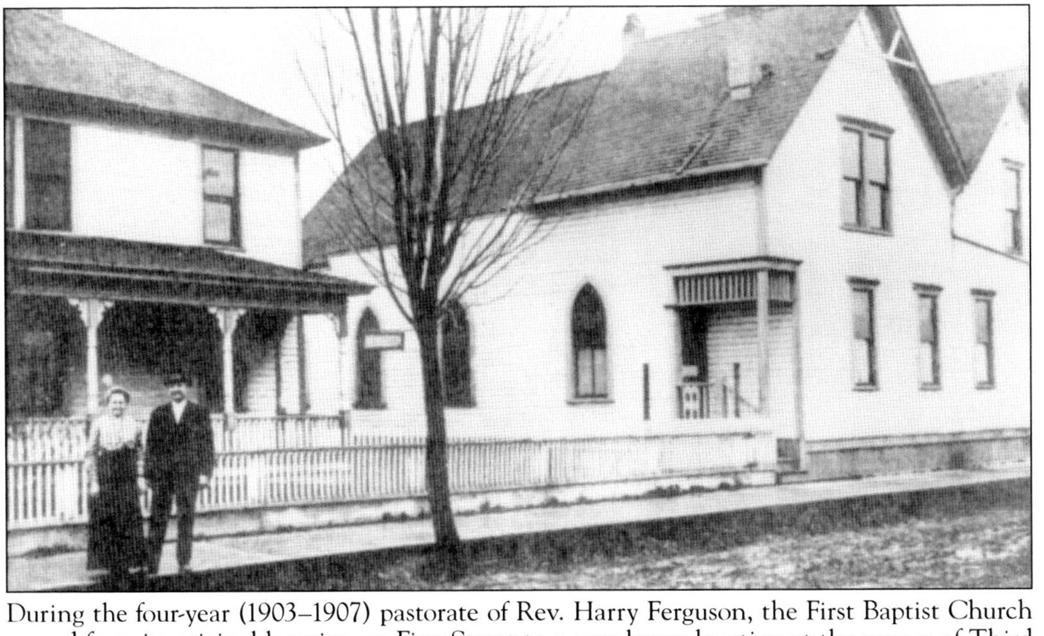

During the four-year (1903–1907) pastorate of Rev. Harry Ferguson, the First Baptist Church moved from its original location on First Street to a new larger location at the corner of Third and Montgomery Streets. A two-story parsonage, seen to the left of the church in this c. 1912 photograph (with Mrs. and Rev. Walter M. Jennings standing in front), was also built. (Courtesy of Mount Vernon First Baptist Church.)

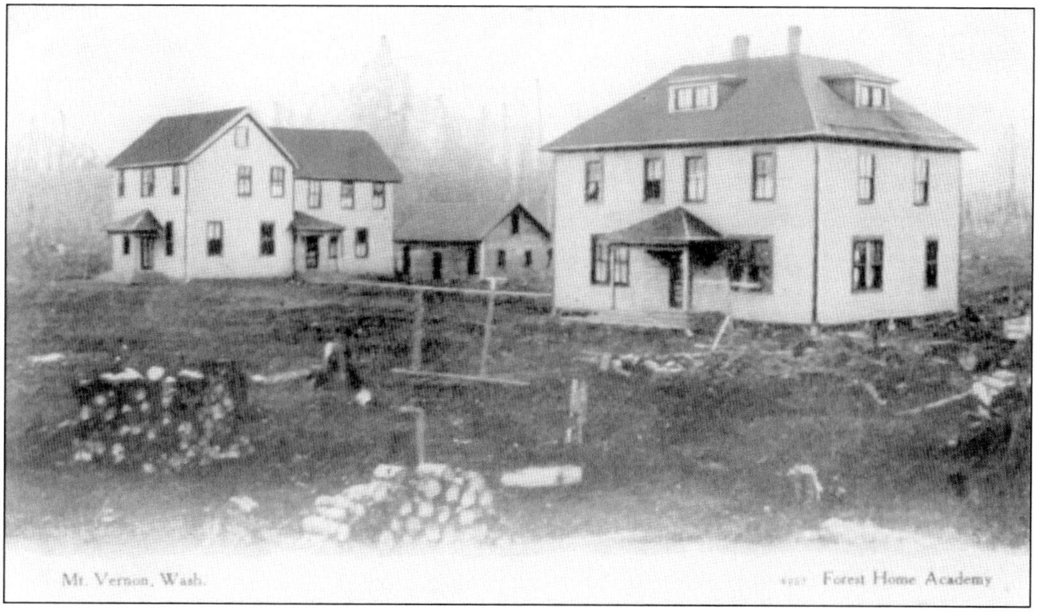

In 1904, the Seventh-day Adventist Church opened Forest Home Industrial Academy, a private school under the leadership of L.I. Stiles, offering industrial education as well as "instruction in all the common English and scientific branches taught in high schools." The campus was dedicated on September 25, 1904. Initially, 26 pupils were enrolled. By 1906, the school had expanded to four teachers and 68 students, despite the annual tuition of $90. The 35-acre campus was located two-and-a-half miles from Mount Vernon; that location is now the corner of College Way and North Waugh Road. Seen below are unidentified members of the school's cornet band. (Above, courtesy of John Kamb Jr.; below, courtesy of the Roger Fox collection.)

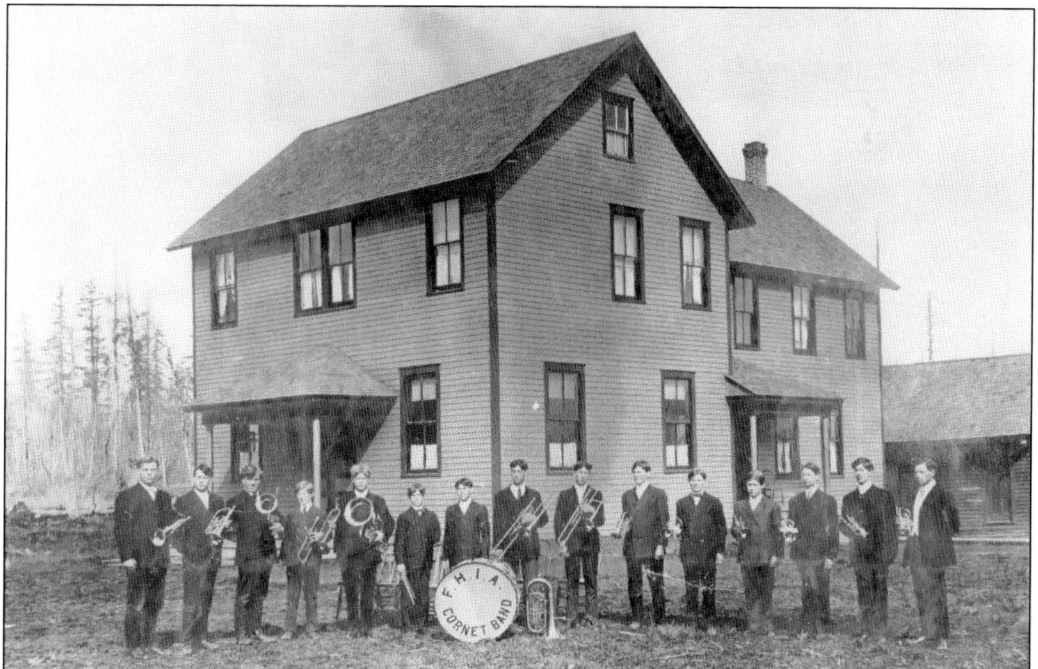

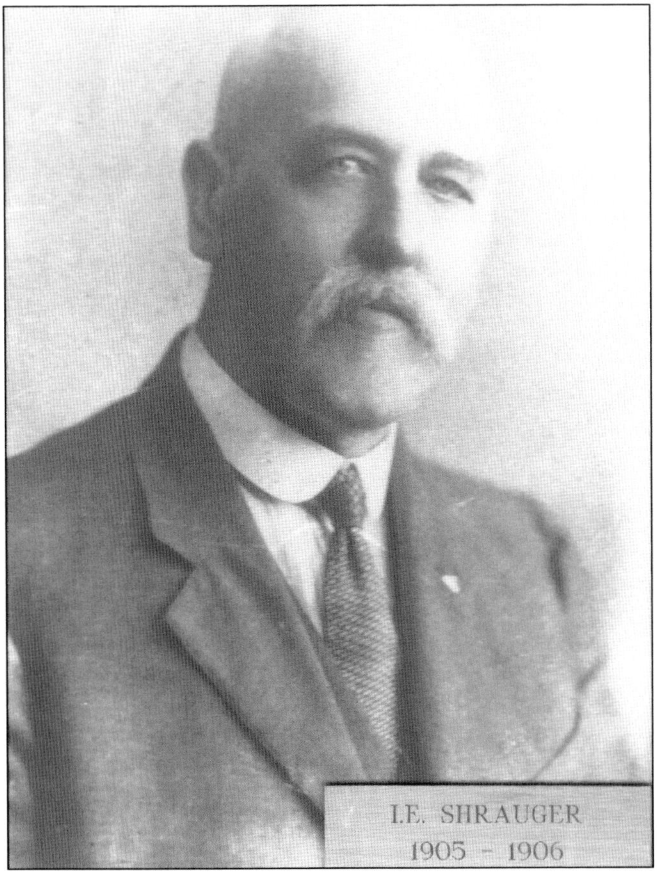

Ira E. Shrauger was the mayor of Mount Vernon from 1905 until 1906. Among other professional and civic duties, he served as president of the Skagit County Bar Association, treasurer for the County Fair Board, and as an officer and active member of the Mount Vernon Commercial Club, the Knights of Pythias, and the Mount Vernon School Board. In 1906, he had the (still-standing) Union Block Building, seen below on the far right, constructed for his law offices. The sign for his office can be seen outside the top left window. Located directly across Pine Street from the old Skagit County Courthouse, the Union Block Building has continually been used as a location for legal offices. (Left, courtesy of the Mount Vernon Mayor's Office; below, courtesy of John Kamb Jr..)

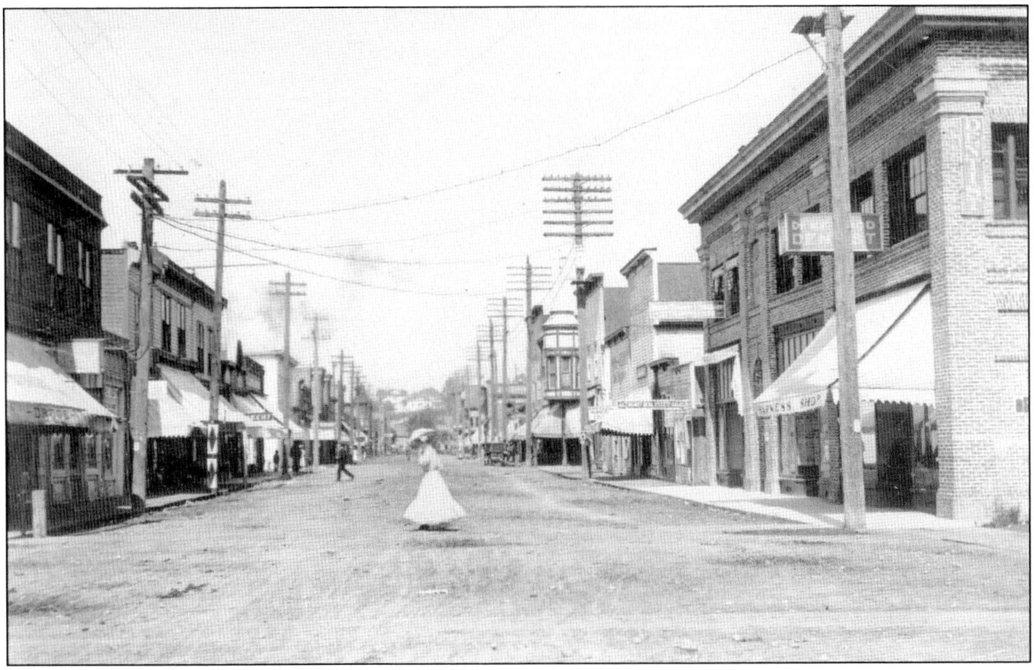

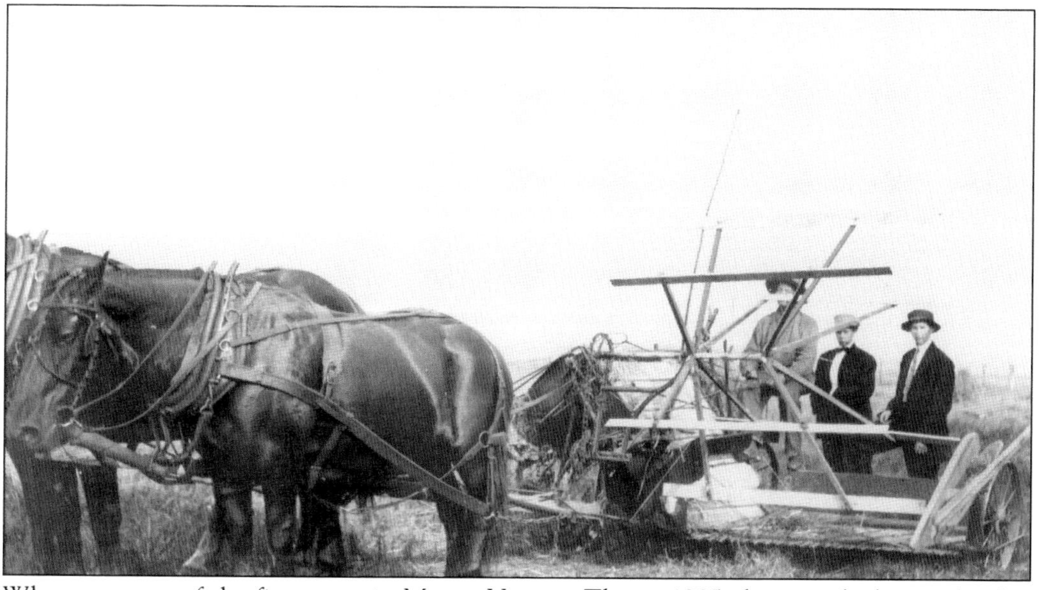

Wheat was one of the first crops in Mount Vernon. This c. 1905 photograph shows a binder. After being cut, or "binded," the wheat would be threshed (a process in which wheat kernels are separated from the rest of the plant), usually in a barn. Around 1910, horse-drawn combines, which did the binding and threshing at the same time in the field, came into use. (Courtesy of John Kamb Jr..)

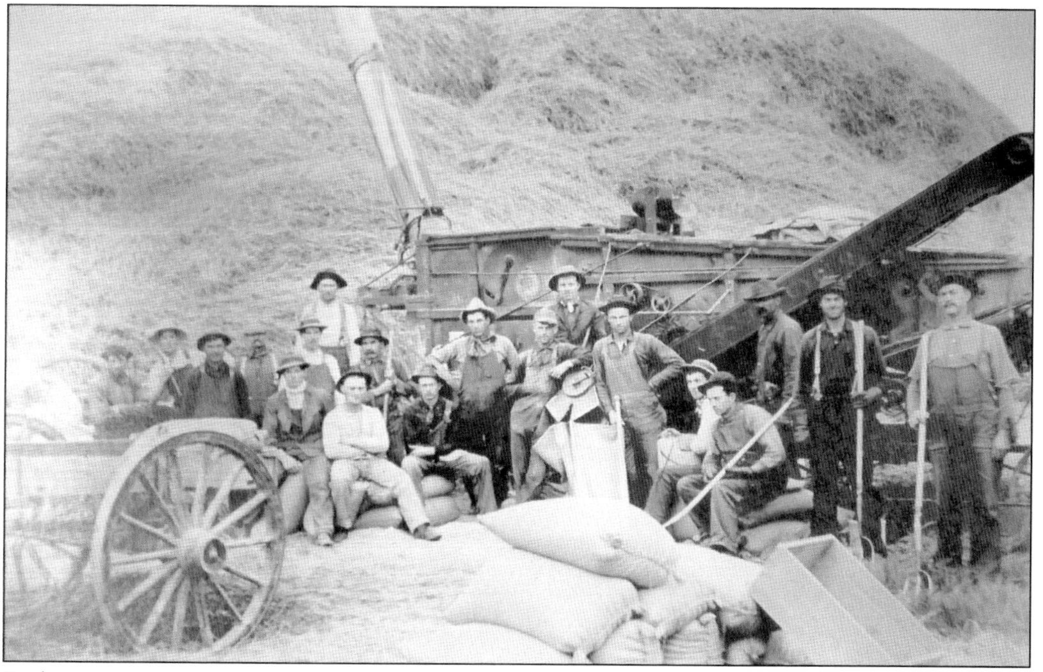

Early on, oats were another main crop in Mount Vernon. At harvest, oats were cut and bundled. Then the threshing rig (like the one seen here) would make the rounds, separating the oats from their stalks. The oats were then put into sacks for transport and storage. Silos were erected along the river, and stern-wheelers transferred oats to Seattle for fueling horse-drawn carriages. (Courtesy of WSU Skagit County Extension.)

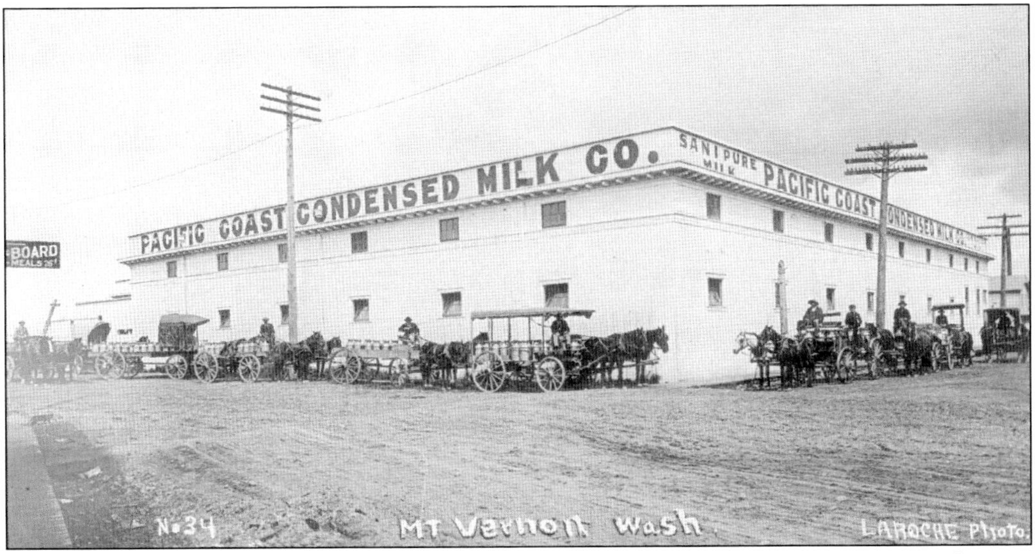

The Pacific Coast Condensed Milk Company began construction on a massive condensery plant in 1906, following active campaigning and negotiating on the part of the Mount Vernon Commercial Club (predecessor of the chamber of commerce). The plant, strategically placed, was accessible by wagon, riverboat, and railroad. In 1907, it was purchased by the Carnation Company of Seattle under the guidance of E.A. Stuart, a pioneer in the processing, evaporation, and canning of milk. The Carnation plant, as well as other milk-processing and condensing plants in the city, contributed substantially to the city's economy. Seen below are unidentified workers in one of the local creameries. (Above, courtesy of John Kamb Jr.; below, courtesy of the Basye family.)

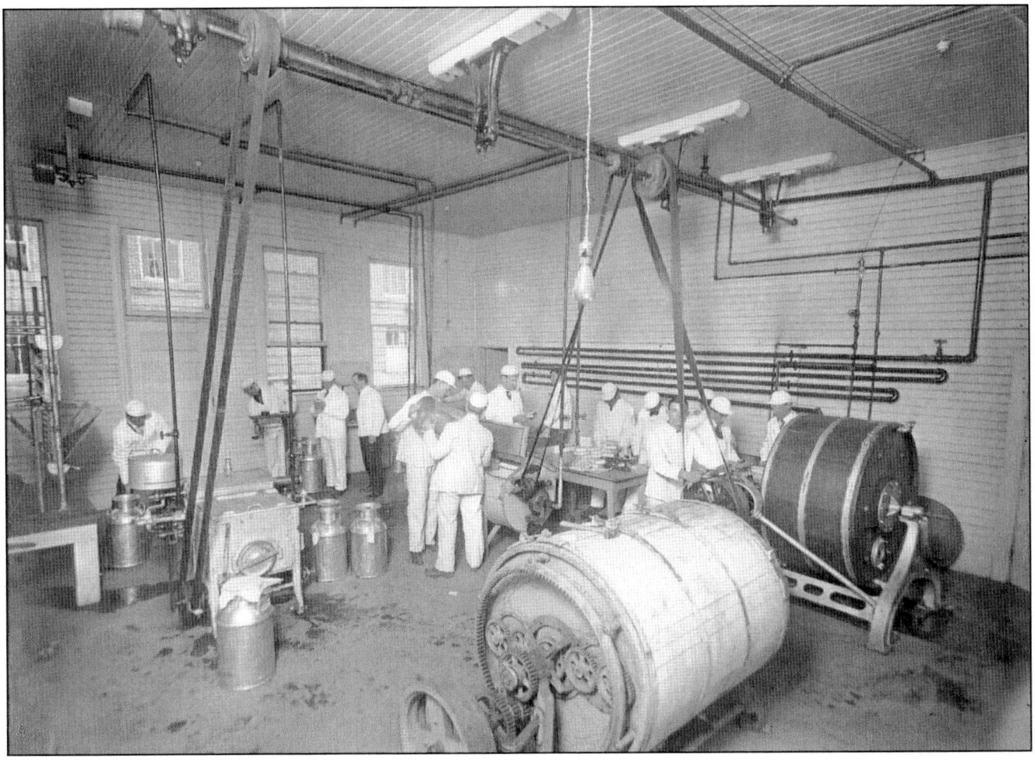

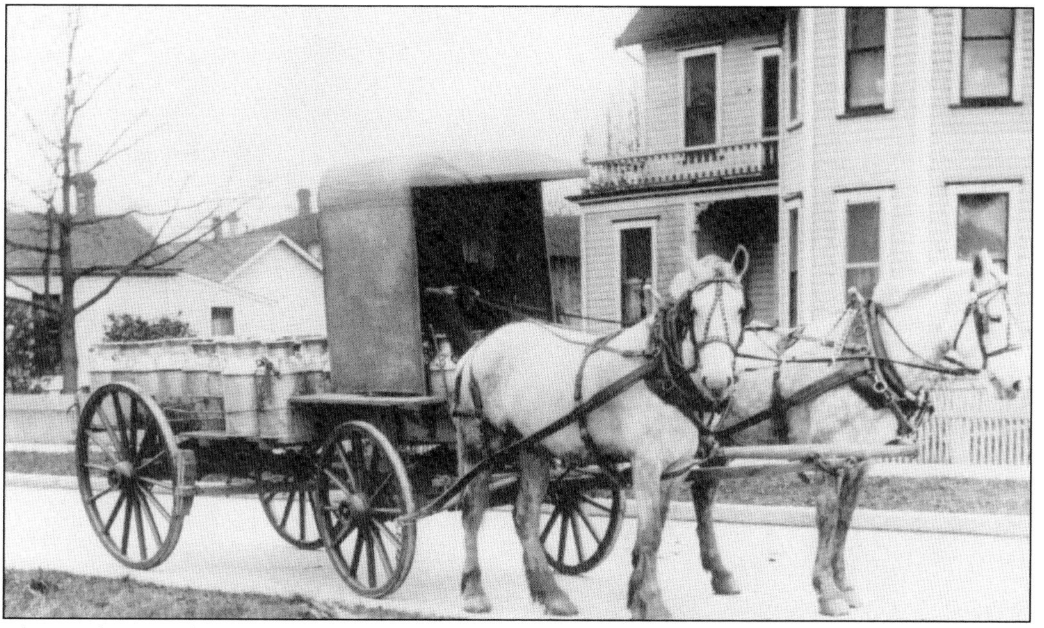

The condensing plants would send wagons on regular routes through the region to collect 10-gallon milk cans from farmers in the morning. The cans were returned, washed, and sterilized each afternoon in time for the evening milking. Seen here is the milk delivery wagon of Iner A. Moe, with his team, Maude and Prince, on Cleveland Street. Moe later upgraded to a delivery truck. (Courtesy of the Roger Fox collection.)

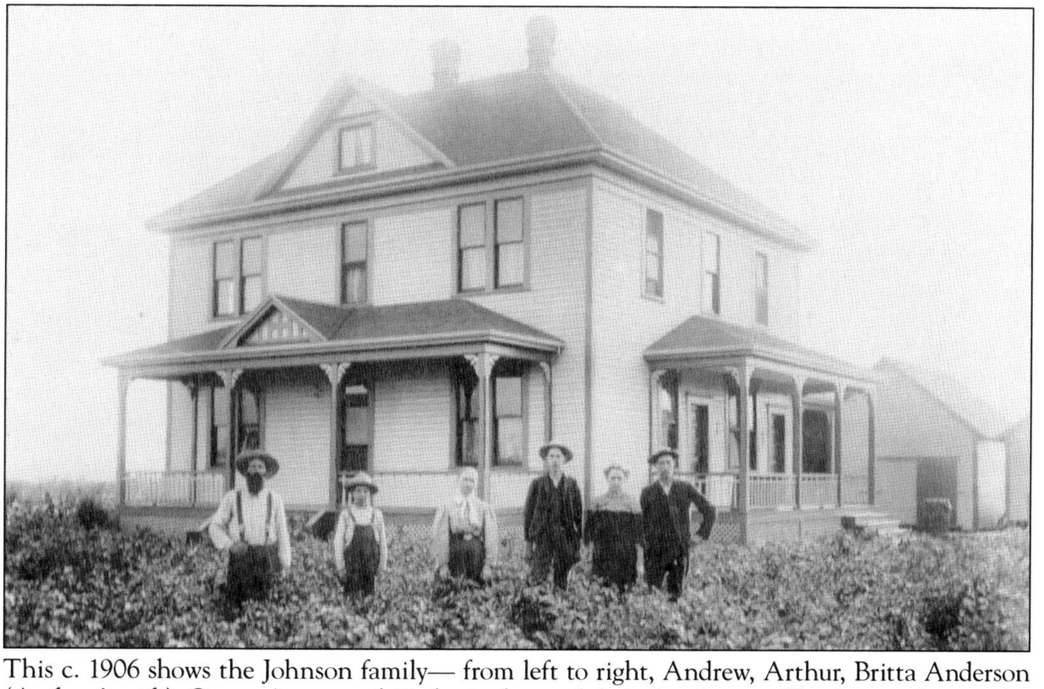

This c. 1906 shows the Johnson family— from left to right, Andrew, Arthur, Britta Anderson (Andrew's wife), Oscar, Anna, and Carl—in front of their farmhouse. The first settlers in the Beaver Marsh area faced a swampy expanse that required clearing and drainage to reveal some of the richest soil in the valley. (Courtesy of the Basye family.)

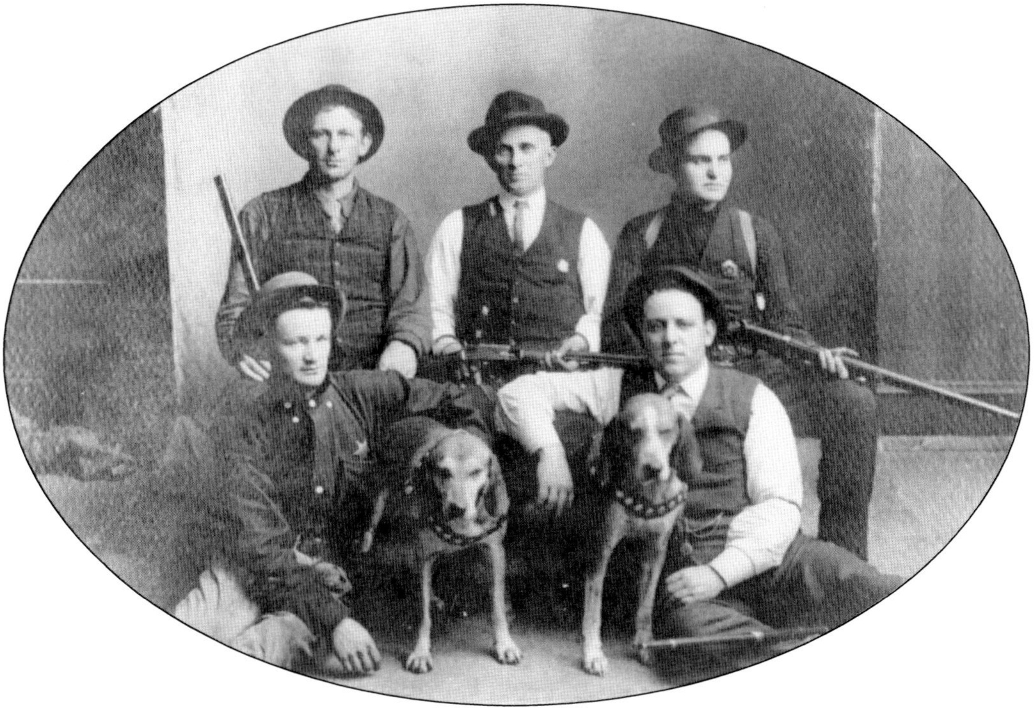

Pictured here are members of an early Mount Vernon Sheriff's Department. They are, from left to right, (first row) unidentified, two hounds, and Billy Bardsley; (second row) Shell Elkins, Charles W. Stevenson (sheriff), and Rick Dunham. Stevenson was sheriff for four terms, between 1908 and 1920, and Ed Wells filled the office for two terms in the middle of that time period (1912–1916). (Courtesy of the Roger Fox collection.)

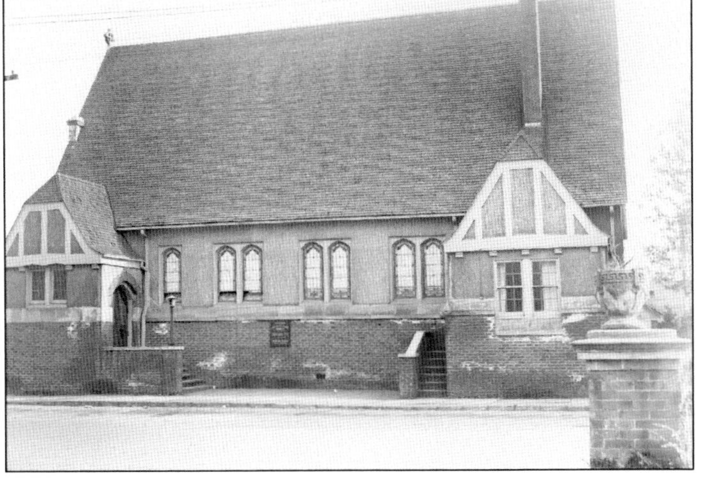

Although it is uncertain when the First Presbyterian Church of Mount Vernon was initiated, in 1907, Rev. Andrew W. Wilson was sent to close the old church because of low membership. Instead, the popular minister assembled an active congregation that built a large new church on the corner of Third and Kincaid Streets in 1921, where they worshiped until 1971. (Photograph by Bill Forman; courtesy of Mark Iverson.)

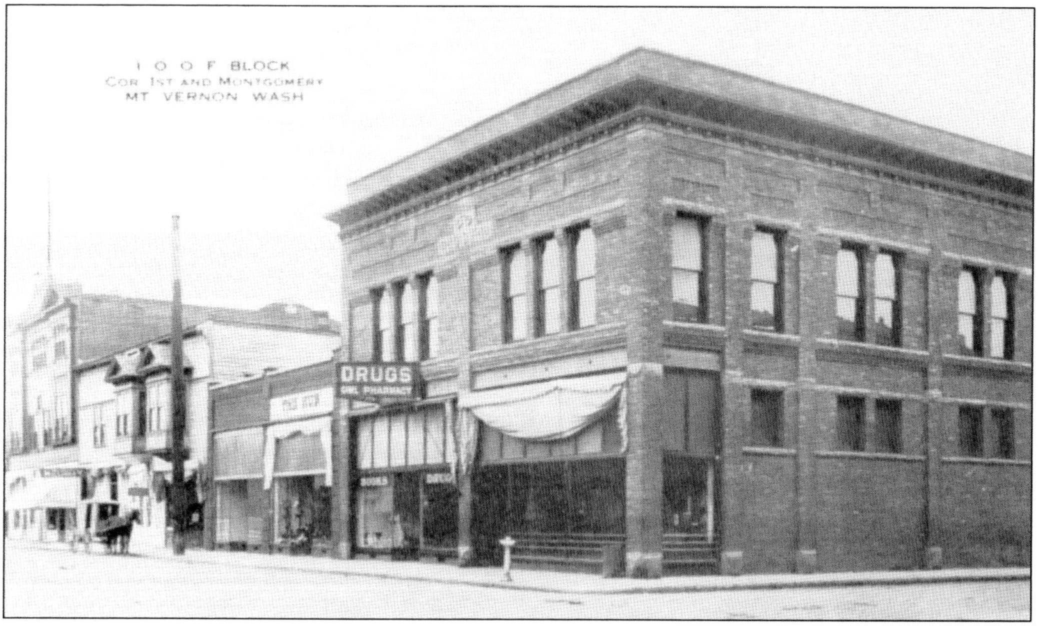

On October 14, 1882, the Independent Order of Odd Fellows became the first of several groups to establish a lodge at Mount Vernon. Many important officials came for the ceremonies, including territorial governor William A. Newell. The original Odd Fellows Lodge building was dedicated on April 27, 1885. Church services were held here by every denomination until each could secure its own building. The large hall served as the first public town hall and as Skagit County's courthouse (1884–1893). In 1908, the original wooden building was moved back, and the brick structure seen above was constructed. The new building contained a spacious ballroom on the second floor and commercial space—including a grocery (seen below) and the post office—on the ground floor. (Both, courtesy of John Kamb Jr..)

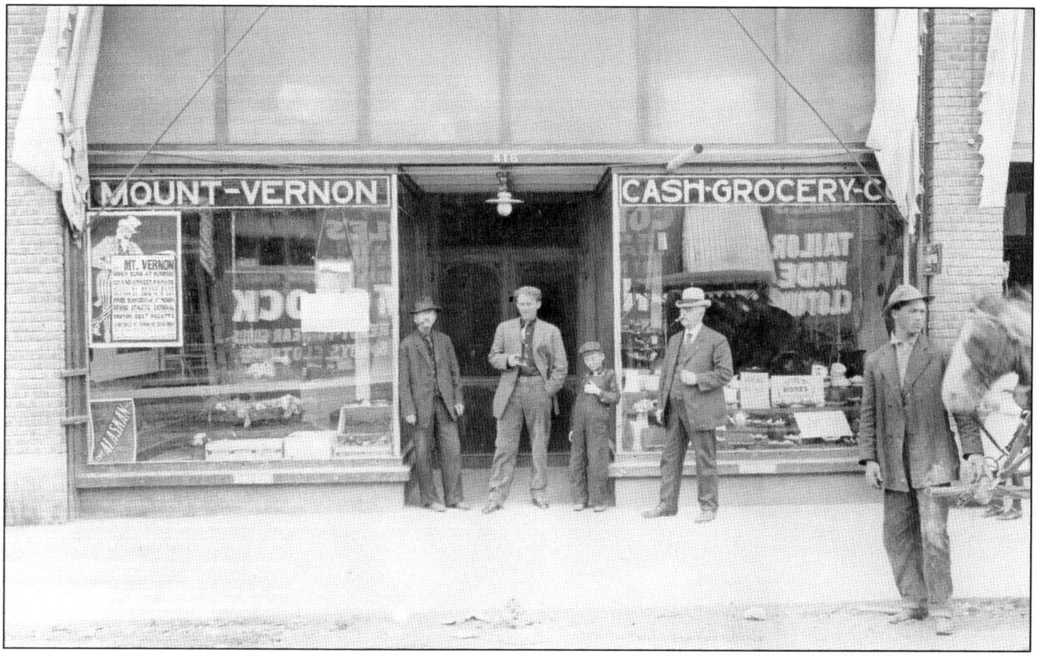

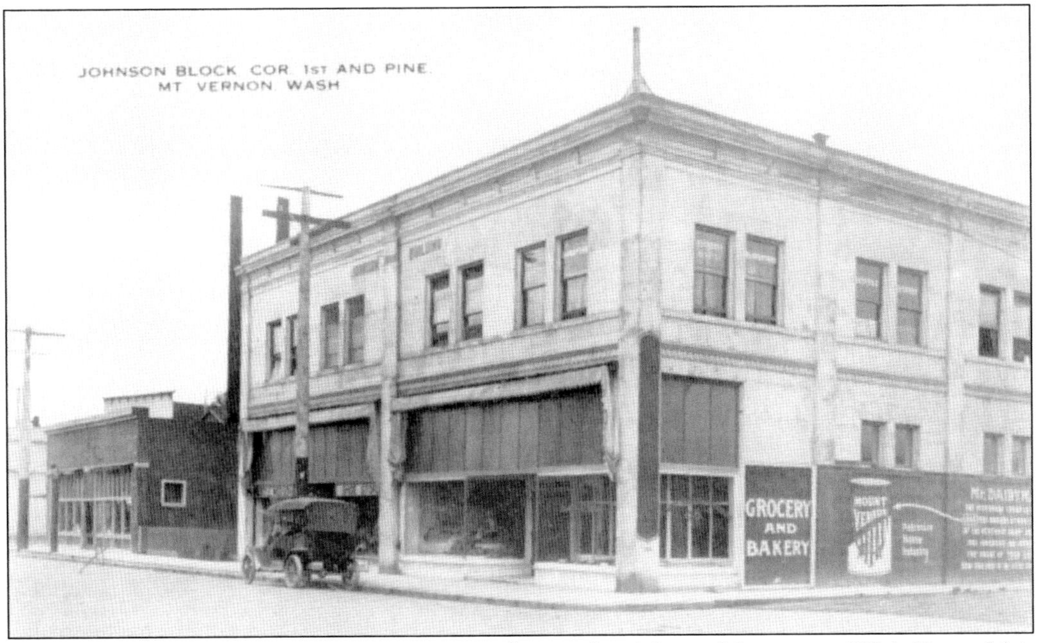

N.B. Johnson came from Sweden in 1880 and moved to Mount Vernon in 1904. He and his three sons, John, Simon, and Fred, started what would become the city's largest grocery firm, N.B. Johnson & Sons Company. The Johnson Building was constructed in 1908 at the corner of First and Pine Streets and also contained Gust E. Werson's boot and shoe store. (Courtesy of John Kamb Jr..)

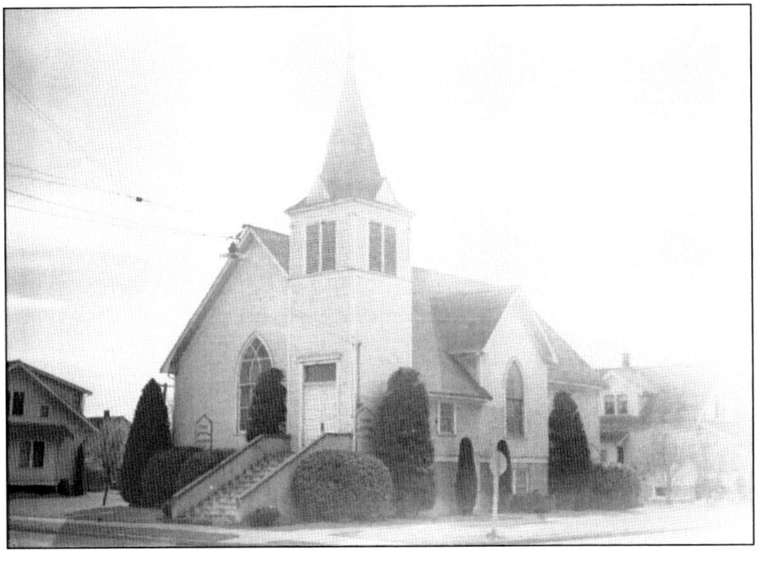

The Norwegian Lutheran Church in Mount Vernon was established in 1908. The church at Second and Section Streets, seen here in the early 1940s, served the congregation until 1973. Over the years, the church became First Lutheran and is now known as First Evangelical Lutheran Church. (Photograph by Bill Forman; courtesy of Mark Iverson.)

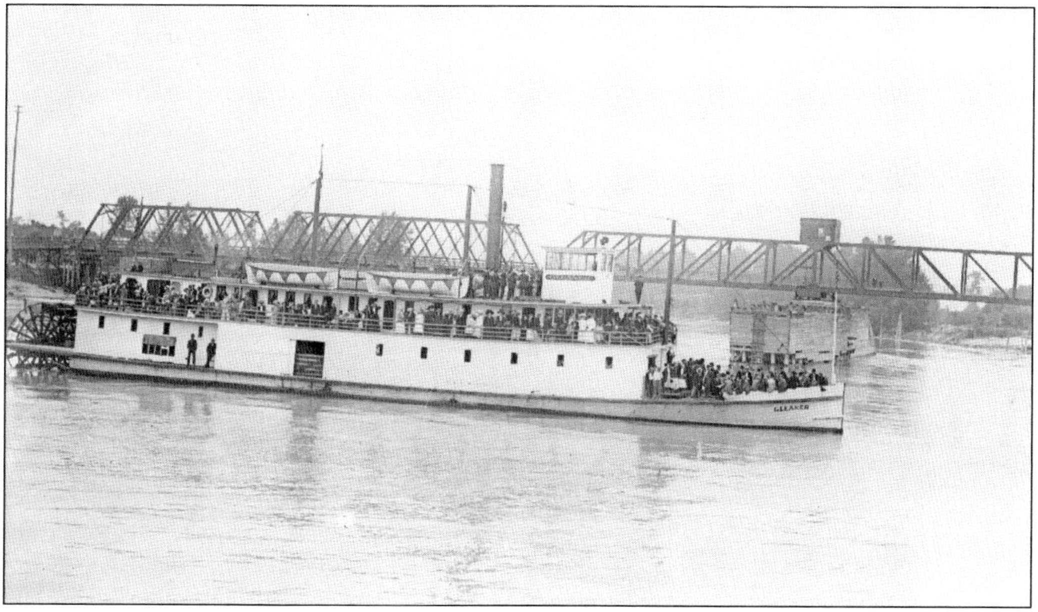

The stern-wheeler *Gleaner*, shown above, was built at Stanwood in 1907 and operated by the Skagit River Navigation & Trading Company, which was founded in 1890. Beginning in 1909, the *Gleaner* would run special Sunday excursions during the summer for up to 200 travelers. The journey began in Mount Vernon at 9:00 a.m., with stops at Skagit City, Fir, Utsalady (on Camano Island), and Rosario Beach (on Fidalgo Island). During the return, local bands, such as the one in the image below, would play music for dancing on the deck. After the *Gleaner* wrecked at the mouth of the Skagit north fork in 1941, it was replaced by the *Skagit Belle*. (Both, courtesy of John Kamb Jr..)

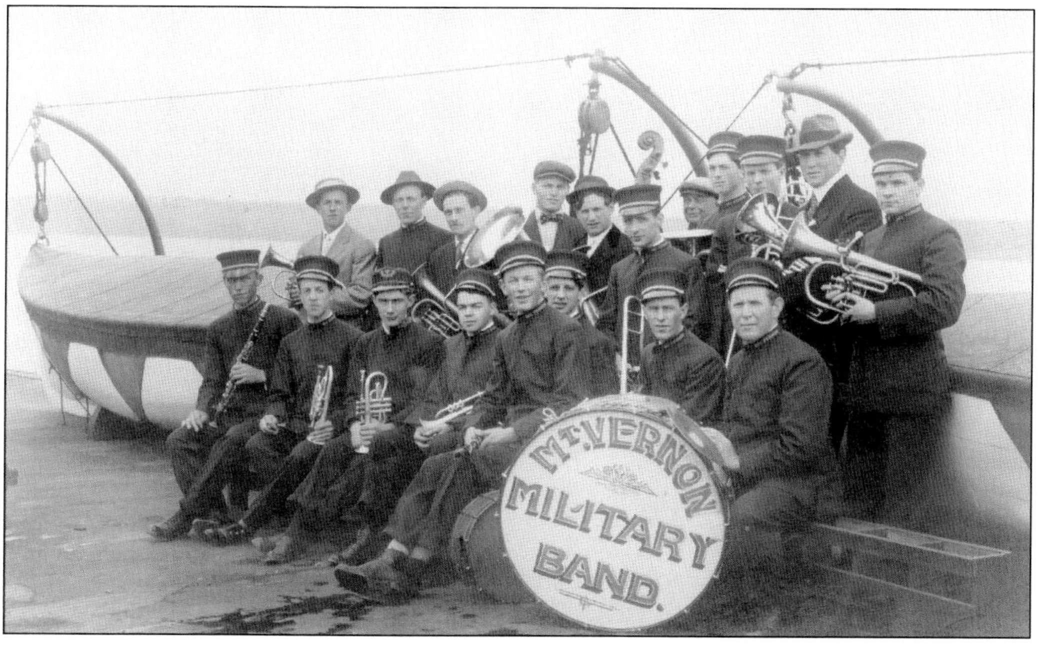

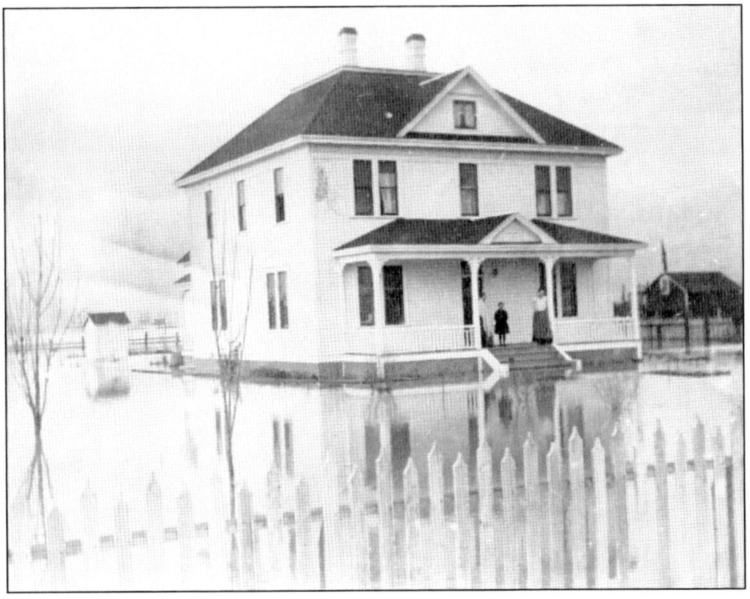

The Johnson family home, located on Beaver Marsh Road, is shown surrounded by floodwaters in November 1909. The flood caused considerable damage to the town's dikes, railways, roads, and property. It also resulted in heavy losses of livestock and valuable farm products. Some farms were covered with sand and gravel, while others were completely washed away. (Courtesy of the Basye family.)

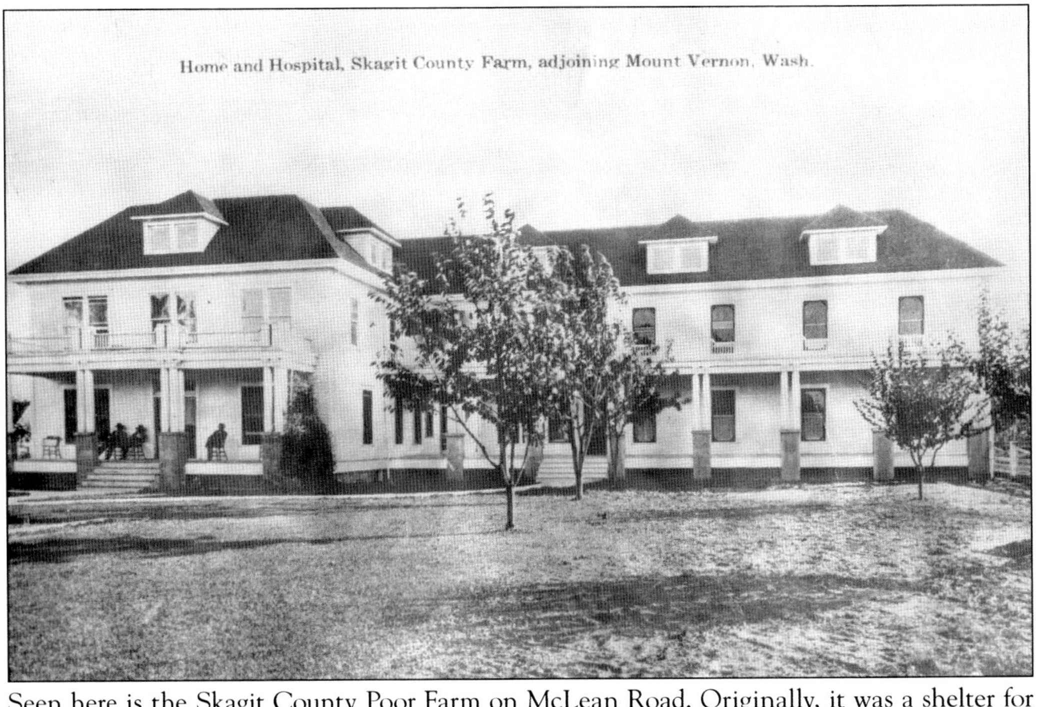

Seen here is the Skagit County Poor Farm on McLean Road. Originally, it was a shelter for men who were homeless or too ill to care for themselves. The facility later housed patients with tuberculosis and evolved into a nursing home for both men and women. According to a special 1910 census of "paupers in almshouses," the home had 11 residents in January 1910. (Courtesy of the Roger Fox collection.)

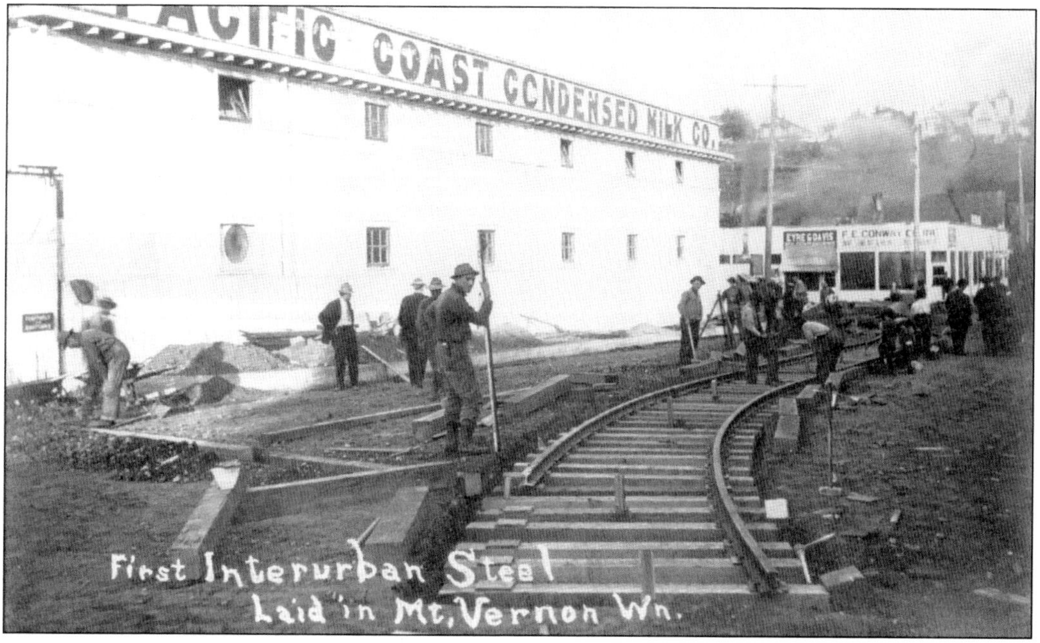

On May 18, 1910, the Bellingham & Skagit Interurban Railway (B&SI) was incorporated, and the company planned tracks to Mount Vernon, Sedro-Woolley, Anacortes, and other points. Three Stone & Webster Management Company investors from Portland, Maine, were the original B&SI incorporators. Above, unidentified workers lay the first interurban tracks in Mount Vernon. The close proximity to area industries, like the Pacific Coast Condensed Milk Company, was essential to the freight aspect of the new railroad. It would cross the Skagit River north of downtown at the area known as Riverside. Seen below is the construction site of the Interurban Bridge. (Both, courtesy of John Kamb Jr..)

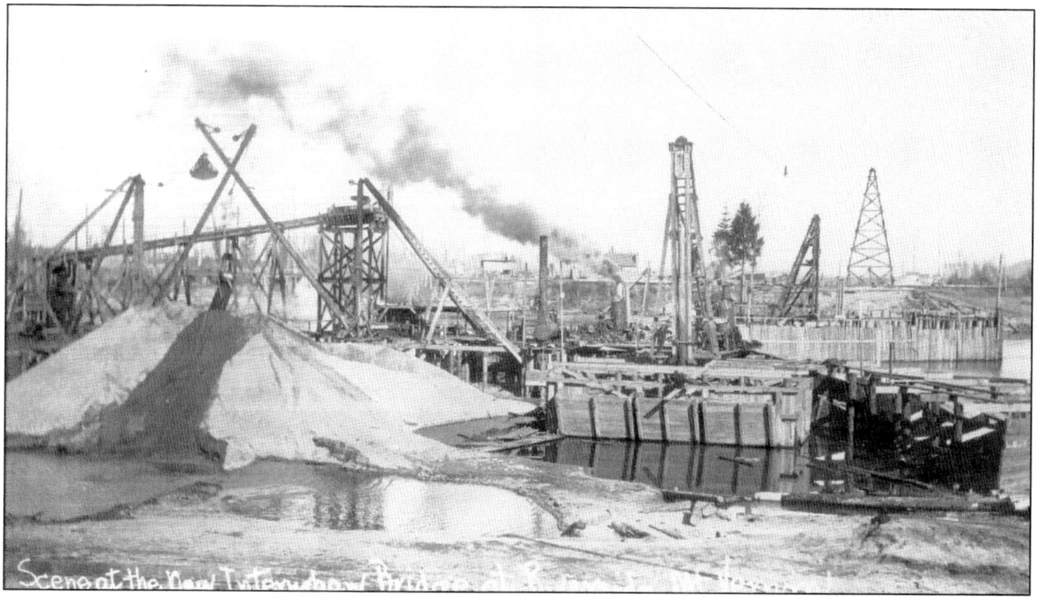

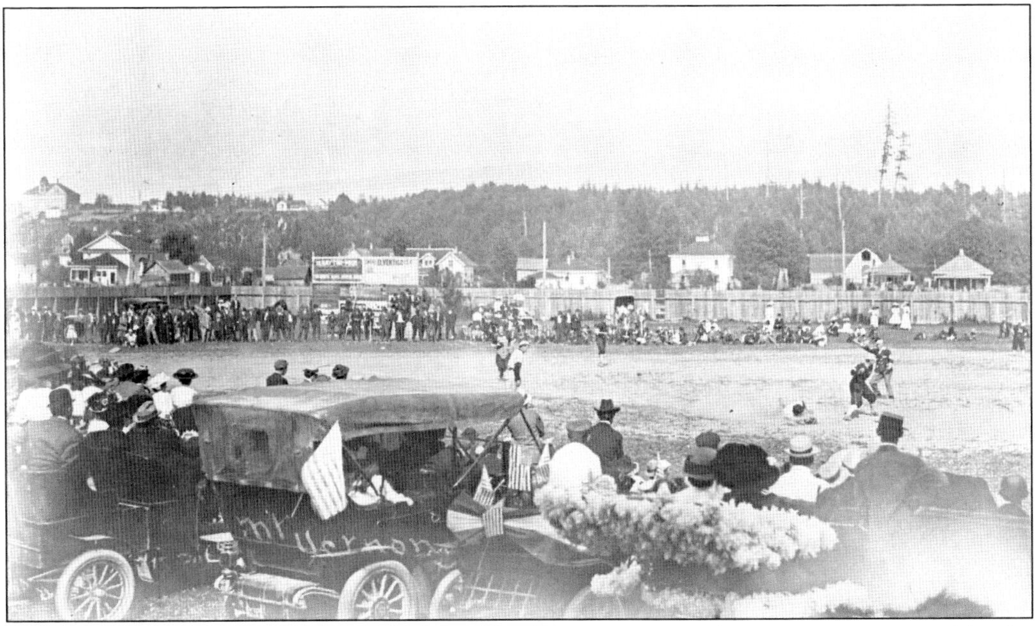

This baseball game was held a few blocks south of downtown, at a location now occupied by the fairgrounds and Sherman Anderson Field. The building on the hill to the left is the old Lincoln School (constructed in 1891). The mix of horse-drawn carriages and automobiles, as well as the American flags with the 45-star pattern, place this picture in the early 1900s. (Courtesy of John Kamb Jr..)

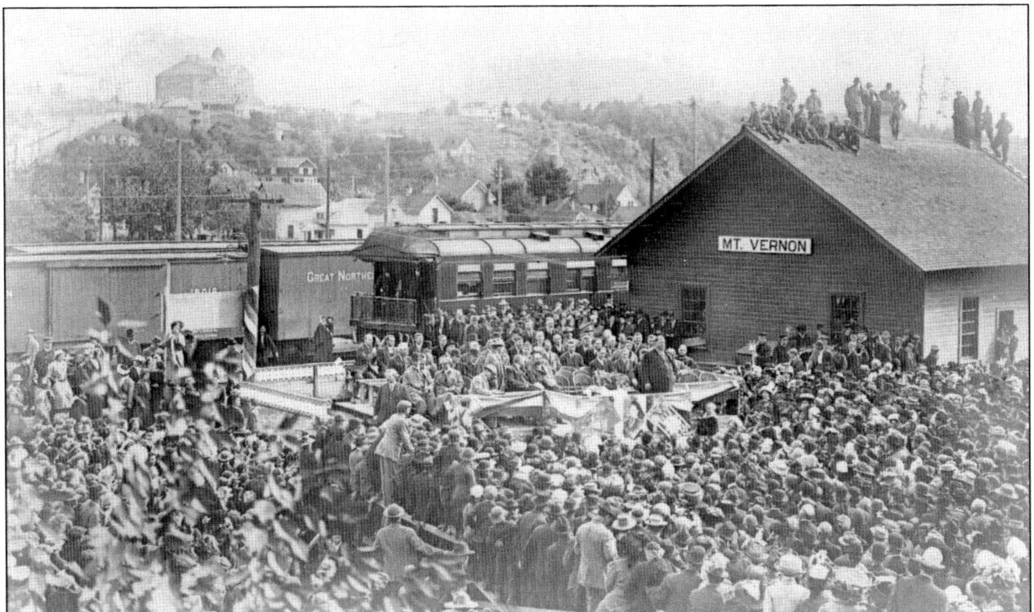

On October 9, 1911, the 27th president of the United States, William Howard Taft, visited Mount Vernon. He stayed at the Hotel Windsor, which, in honor of his visit, would be renamed the President Hotel following a reconstruction in 1927. It is rumored that, while there, the portly president got stuck in a bathtub. (Courtesy of John Kamb Jr..)

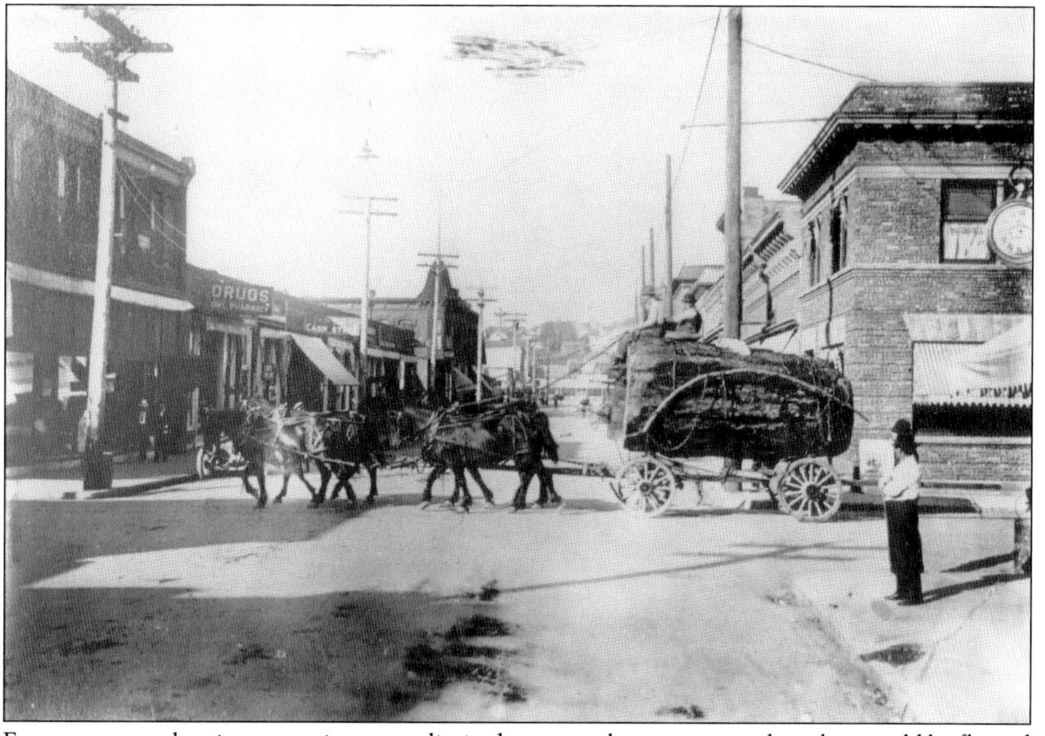

For many years, logging operations were limited to areas close to water, where logs could be floated to mills. As more land was logged, more transportation was required. Large logs were often hauled through downtown Mount Vernon and taken to mills or dumped in the Skagit River for transport. Seen above at the corner of First and Gates Streets, Ernie Rothrock and Mert Holly bring logs from Lincoln Hill to the river. Below is the dump ramp into the Skagit River, which was located on Gates and Front Streets. Holding the reins is Ernie Rothrock, along with his brother Ray, Bill Lee, and an unidentified man. (Both, courtesy of the Roger Fox collection.)

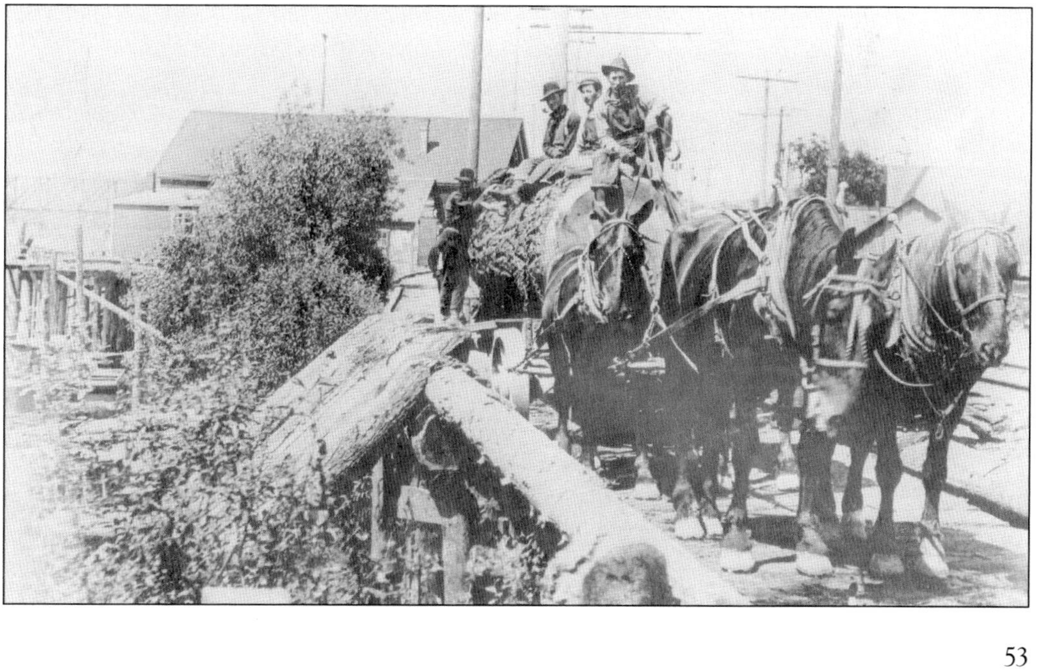

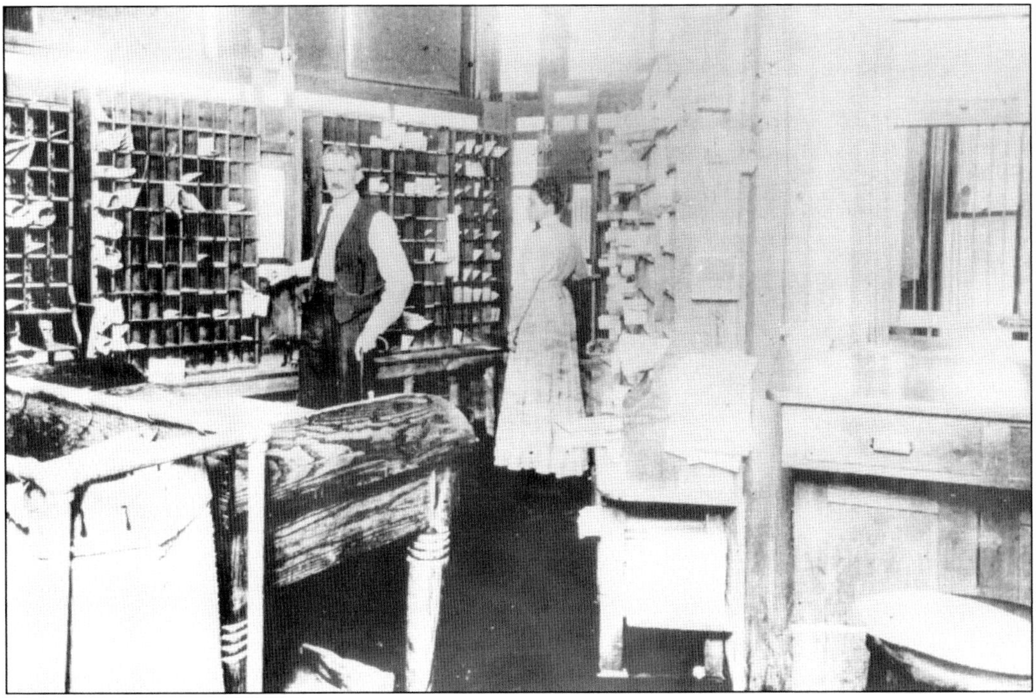

George E. Hartson was married to Matilda Gates, daughter of Jasper Gates, in 1879. He was appointed postmaster of Mount Vernon in 1902 and is pictured above in the post office. At that time (and for many years) the post office was located in the Odd Fellows Lodge at the corner of First and Montgomery Streets. Pictured below, outside the lodge, are mail carriers with their delivery carriages in 1914. The carriers are, from left to right, Billie Stevens, Charley Warren, Virgil Blackburn, Harry White, and Bill Parker. (Above, courtesy of the Roger Fox collection; below, courtesy of John Kamb Jr..)

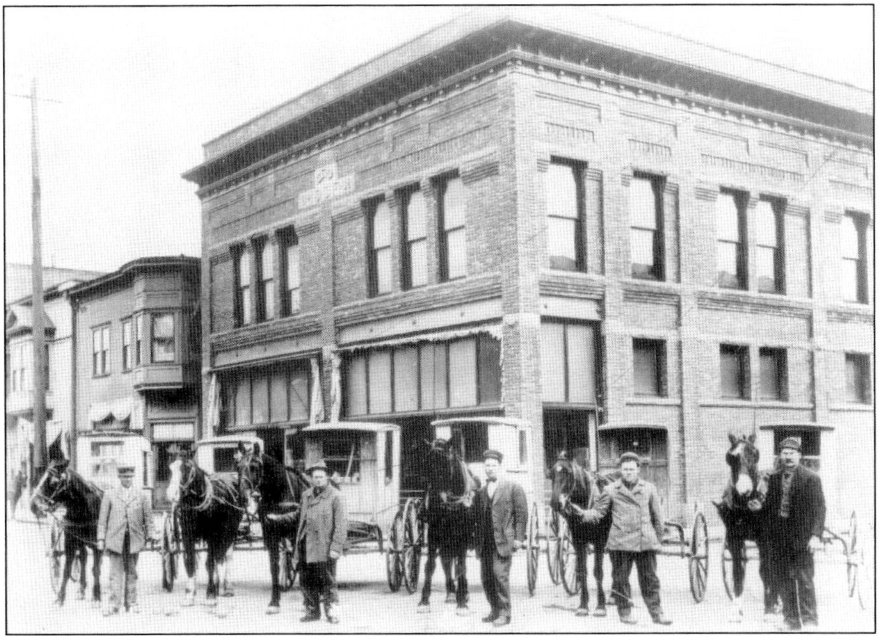

Five

Progression and Preservation 1912–1929

While river and rail traffic had increased the flow of settlers and visitors alike, the introduction of the Bellingham & Skagit Railway's interurban railroad line was yet another boon. In 1912, Mount Vernon became the southern terminal of the electric rail system that carried passengers and freight between Bellingham, Mount Vernon, Burlington, and Sedro-Woolley.

Municipal expansion was also back on track. The first city library was established in 1916. In 1922, a new courthouse (still in use today), designed by Bellingham architect Thornton F. Doan, was built. That same year, the Windsor Hotel expanded by adding two more stories, becoming the town's first "skyscraper." In 1925, the Knights of Pythias Lodge was erected downtown. The Lincoln Theater opened in 1926. In the late 1920s, Second Street was incorporated as part of US Route 99 ("the Pacific Highway"), and the viaduct was upgraded from wood to concrete in 1929.

In the spring and summer of 1920, Mount Vernon and the surrounding community came together to establish one central high school. Union High School District One was created out of 14 separate school districts—an amalgamation that had begun in 1899 when Riverside District Number 38 united with Mount Vernon District Number 10. During July, at the cost of $16,200, twelve and a half acres were purchased as the future site of Union High School. In October, the taxpayers overwhelmingly voted bonds for $225,000 to fund the construction, which was completed in 1922. The campus later included Mount Vernon Junior College, established in 1924, which would eventually become Skagit Valley College.

As logging and development encroached on the area, many citizens recognized the need for preservation, and a city park was proposed in 1923. On March 19, 1924, Hillcrest Park was created when the city council voted to purchase 29 acres of the McGinnitie property on the hill just south of the city limits, at a cost of $6,000. Meanwhile, the Civic Club and other women's groups worked to turn a 240-acre land donation from the English Lumber Company into a 480-acre park at Little Mountain.

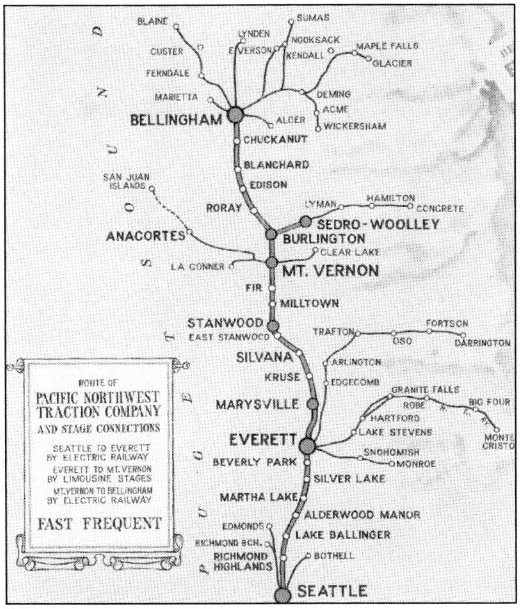

The Pacific Northwest Traction Company was a subsidiary of Stone & Webster Management Company. Founded in Boston by engineer-entrepreneurs Charles Stone and Edwin Webster, it specialized in development of urban utilities and transportation systems across the nation. Pacific Northwest Traction would eventually connect Seattle, Everett, Mount Vernon, Snohomish, and Bellingham, creating the state's largest interurban railroad. (Courtesy of John Kamb Jr..)

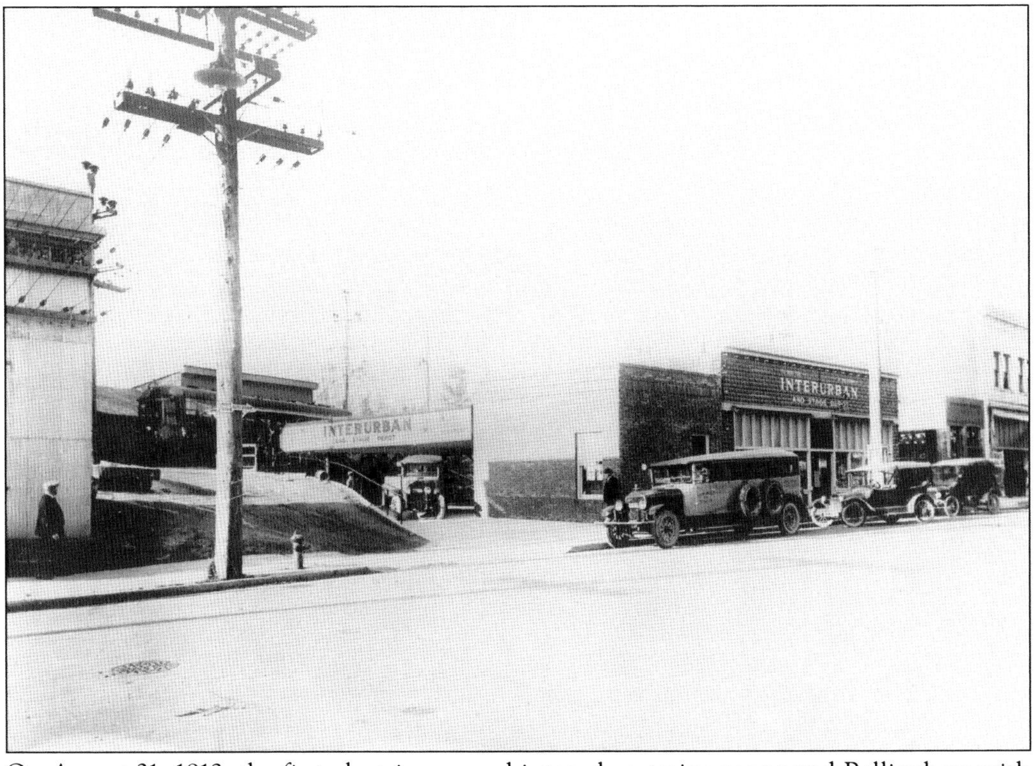

On August 31, 1912, the first electric-powered interurban trains connected Bellingham with Mount Vernon and Sedro-Woolley. The missing rail link, between Mount Vernon and Everett, was powered by motor coaches, like those pictured waiting outside the interurban station. The station, located at the corner of First and Kincaid Streets, was the site of the first electric plant in all of Skagit County, established by W.H. Franklin in the summer of 1891. (Courtesy of the Roger Fox collection.)

As displayed on the outer cover of an interurban flier, its mottos were "fast" and "frequent." In addition to numerous passenger routes, the interurban lines also hauled freight. The construction of the Bellingham and Skagit Railway (which was renamed Pacific Northwest Traction Company in early 1912) represented an attempt to funnel the agricultural bounty of the Skagit Valley through Bellingham, the nearest seaport and major regional railroad center. (Courtesy of John Kamb Jr..)

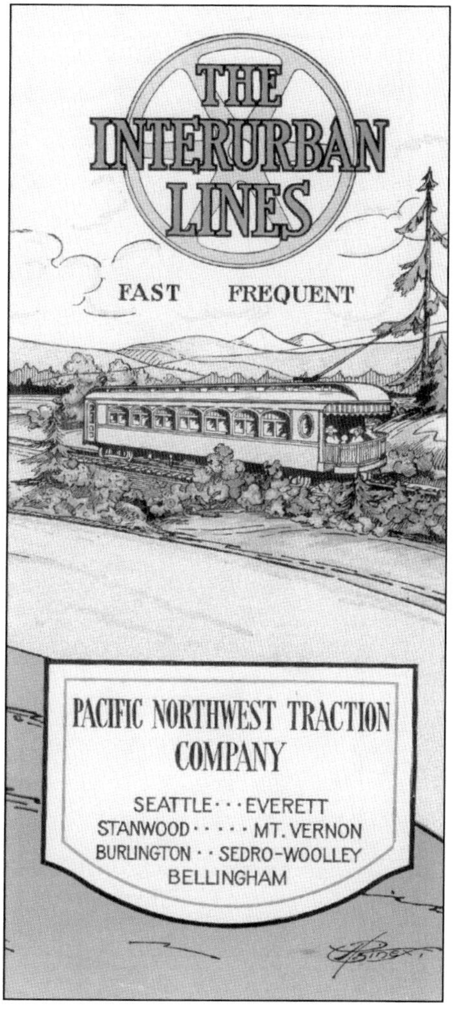

INTERURBAN

TIME CARD

Effective May 12, 1913

LEAVE SEDRO-WOOLLEY

7:00 a.m.	2:30 p.m.	9:40 p.m.
8:10 a.m.	**3:40** p.m.	10:40 p.m.
9:30 a.m.	5:00 p.m.	**11:40** p.m.
10:40 a.m.	**6:10** p.m.	12:10 a.m.
12:00 noon	7:30 p.m.	
1:10 p.m.	**8:40** p.m.	

Black-face figures denote cars connecting at Burlington for Bellingham.

LEAVE MT. VERNON

5:45 a.m.	12:40 p.m.	6:50 p.m.
7:40 a.m.	1:50 p.m.	8:10 p.m.
8:50 a.m.	3:10 p.m.	9:10 p.m.
10:10 a.m.	4:20 p.m.	10:10 p.m.
11:20 a.m.	5:40 p.m.	11:10 p.m.

LEAVE BURLINGTON FOR SEDRO-WOOLLEY

6:20 a.m.	2:05 p.m.	9:25 p.m.
7:55 a.m.	3:25 p.m.	10:25 p.m.
9:05 a.m.	4:35 p.m.	11:25 p.m.
10:25 a.m.	5:55 p.m.	11:55 p.m.
11:35 a.m.	7:05 p.m.	
12:55 p.m.	8:25 p.m.	

LEAVE BELLINGHAM

5:00 a.m.	9:30 a.m.	5:00 p.m.
6:15 a.m.	12:00 noon	7:30 p.m.
7:00 a.m.	2:30 p.m.	11:00 p.m.

The "frequency" of the interurban is confirmed by the 1913 timetable pictured at left. Failure to connect the Bellingham-Mount Vernon segment with the Seattle-Everett portion of the new line (because of World War I) caused a lack of capital, dooming the line. Over time, people began traveling more by car and less by train. The railway was no longer profitable, and the tracks started failing. The railroad ceased operation in 1930. (Courtesy of John Kamb Jr..)

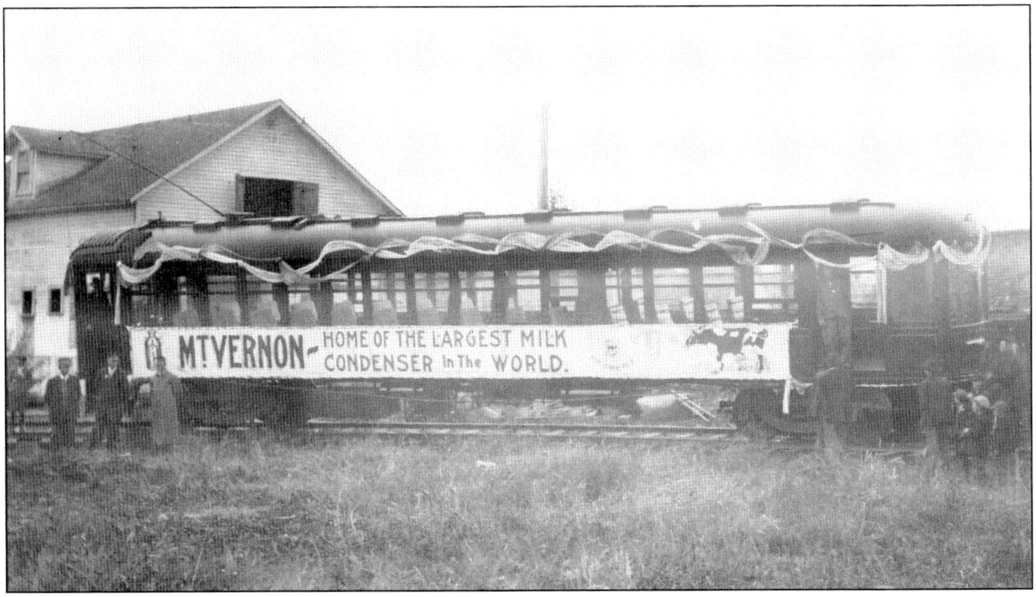

On the interurban, local runs were just a nickel, affording travelers headed north to Bellingham an exhilarating ride over the bay, under the steep cliffs of Chuckanut, and atop a rickety wooden trestle. Pictured above is a Mount Vernon interurban train car. (Courtesy of John Kamb Jr..)

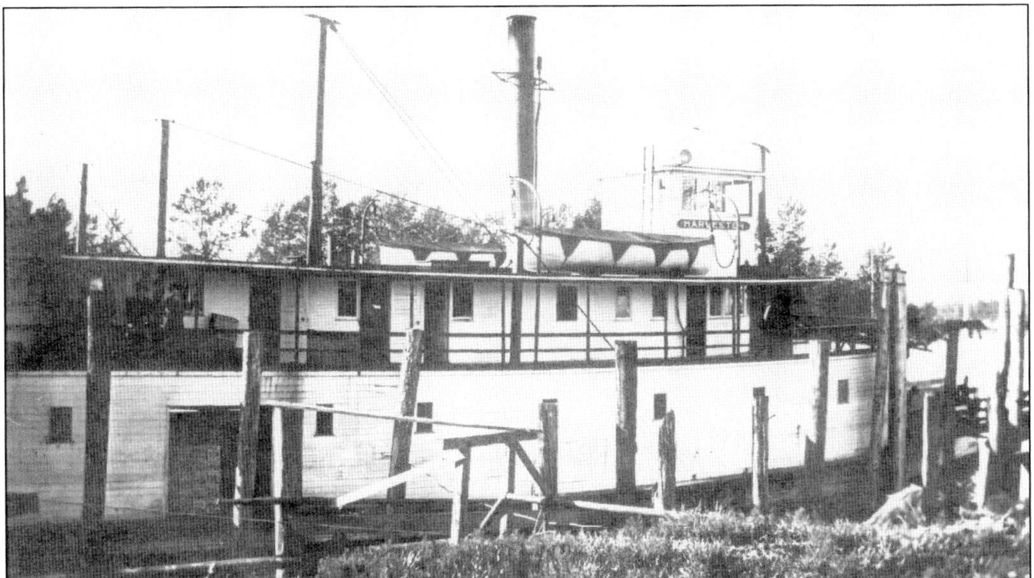

The *Harvester* was built in Stanwood in 1912, using machinery from the former *Skagit Queen*, for the Skagit River Navigation Company. The stern-wheeler traveled from Seattle to Mount Vernon three times a week carrying freight and up to 30 passengers. On October 22, 1934, the *Harvester* sank after being struck by the SS *President Madison,* which had been blown off course by a gale. The next day, she was replaced by the *Gleaner.* (Courtesy of John Kamb Jr..)

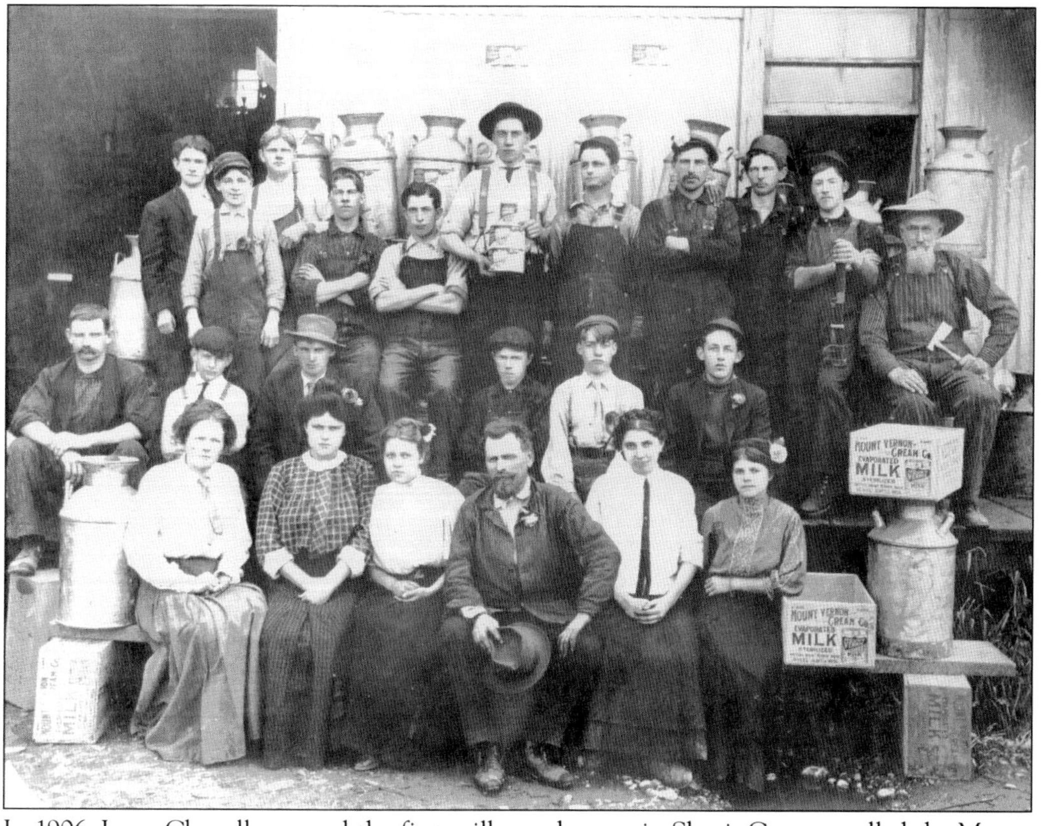

In 1906, Leon Chevally opened the first milk condensery in Skagit County, called the Mount Vernon Cream Company (not to be confused with the earlier Mount Vernon Creamery, which manufactured mainly butter). The process of evaporating and canning milk was being refined at the time, and some believed condensed milk would replace fresh milk in the American market. The company manufactured Mount Vernon Evaporated Milk, a profitable brand that had additional processing plants in Ferndale, Seattle, and Tenino. In 1916, the Mount Vernon Cream Company was sold to the company's main competitor, the Carnation Company, who shut down the condensery in September 1917. The image above is of a group of unidentified Mount Vernon Cream Company employees. The one below shows a parade entry for the company. (Above, courtesy of the Roger Fox collection; below, courtesy of John Kamb Jr..)

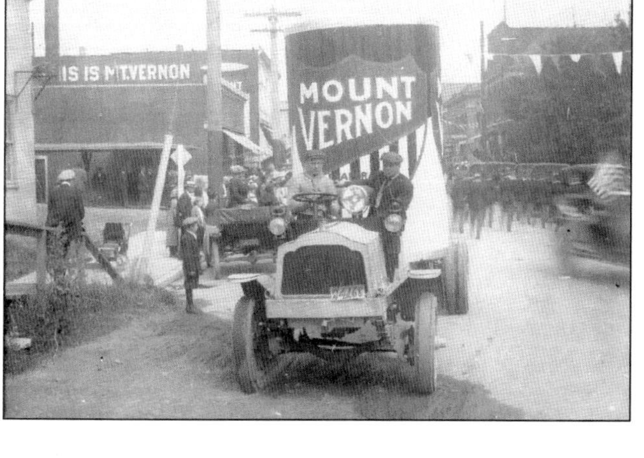

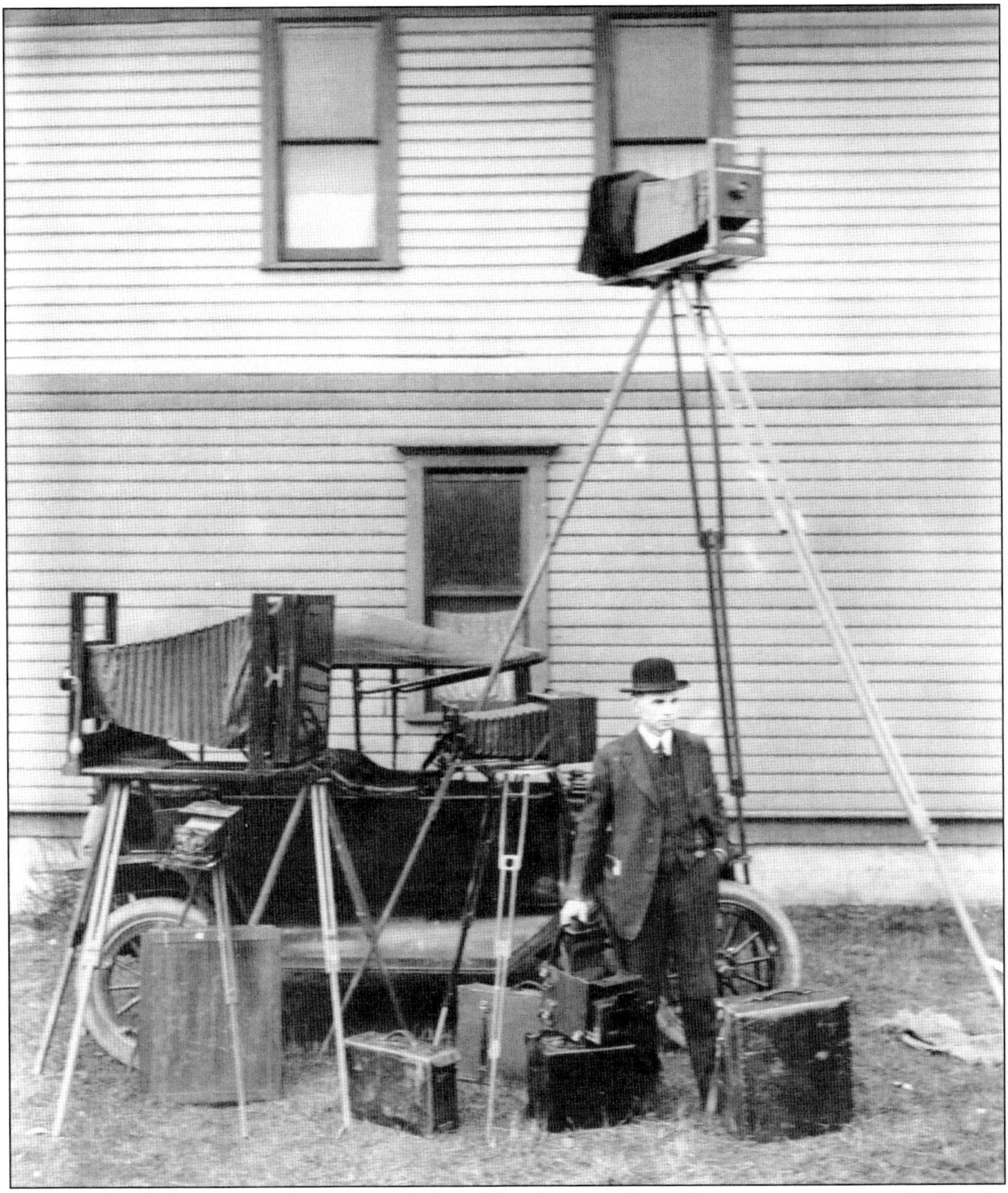

Darius Kinsey was the most important and prolific photographer of logging activities in the Pacific Northwest from the turn of the century until the 1940s. He was born in Maryville, Missouri, in 1869 and moved to Snoqualmie, Washington, where he took up photography in 1890. He worked as a traveling photographer for several years until marrying Tabitha May Pritts in 1896. The following year, the couple set up a studio in Sedro-Woolley. After moving to Seattle in 1906, Darius gave up studio work and focused exclusively on the lumber industry and scenic photography. Tabitha would develop the negatives and send the prints back to the logging camps to be sold to the loggers. Kinsey's equipment was large and cumbersome, as seen in the photograph, and the cameras used fragile glass plates. The smaller camera on the left was used to take stereopticon images. In 1940, Kinsey broke several ribs in a fall from a stump, ending his photographic career. He died five years later in 1945. (Courtesy of the Skagit County Auditor's Office.)

Darius Kinsey, who has been called "the Ansel Adams of the logging industry," took this photograph of an unidentified assistant resting with a heavy case of the fragile 20-by-24-inch glass-plate negatives on his back. The high quality and detail of Kinsey's prints was achieved by making contact prints from original glass negatives. (Courtesy of the Skagit County Auditor's Office.)

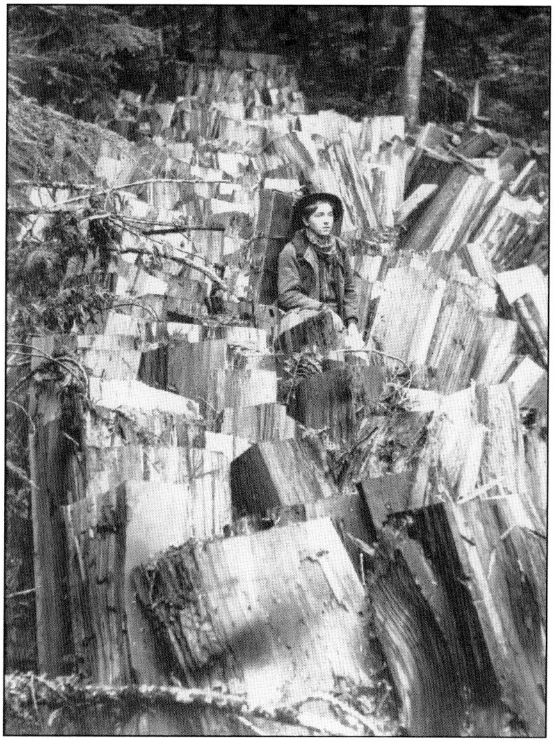

Logging camps were set up so that logs could be easily conveyed to water for transport. Teams of oxen or horses, as seen here, were used to haul a series of four to six logs fastened together, called a "turn." A skid greaser would walk ahead of the team, swabbing the skids with oil so the logs would move easier. (Photograph by Darius Kinsey; courtesy of John Kamb Jr..)

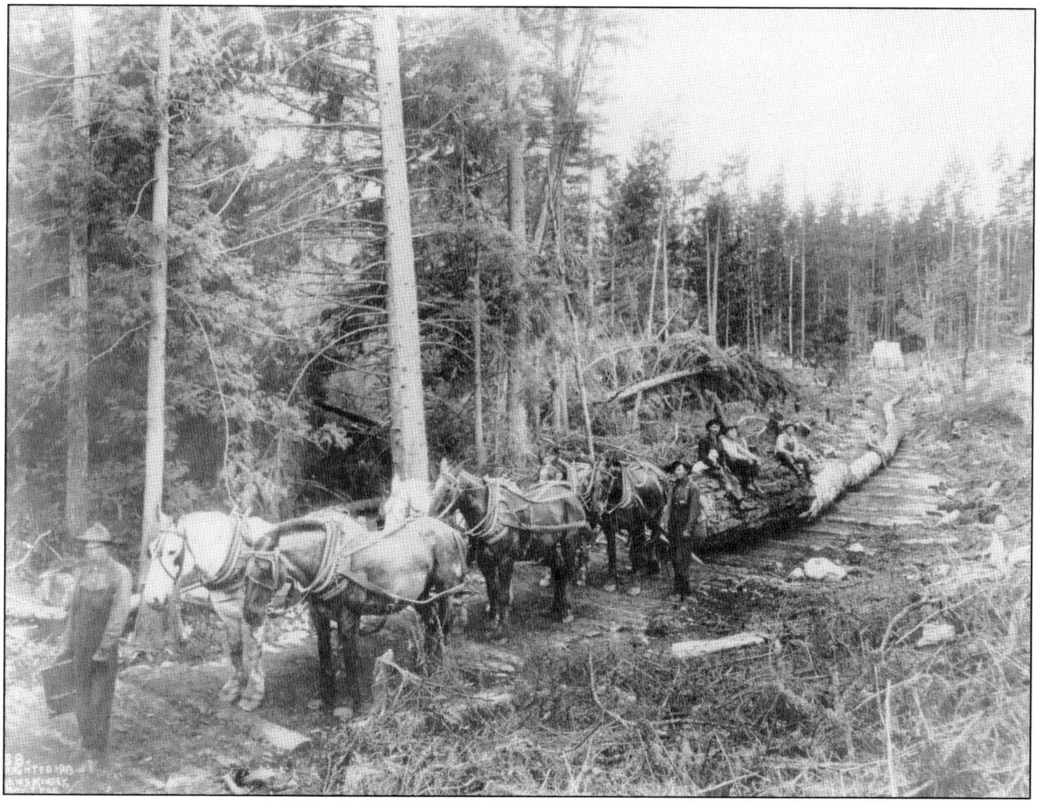

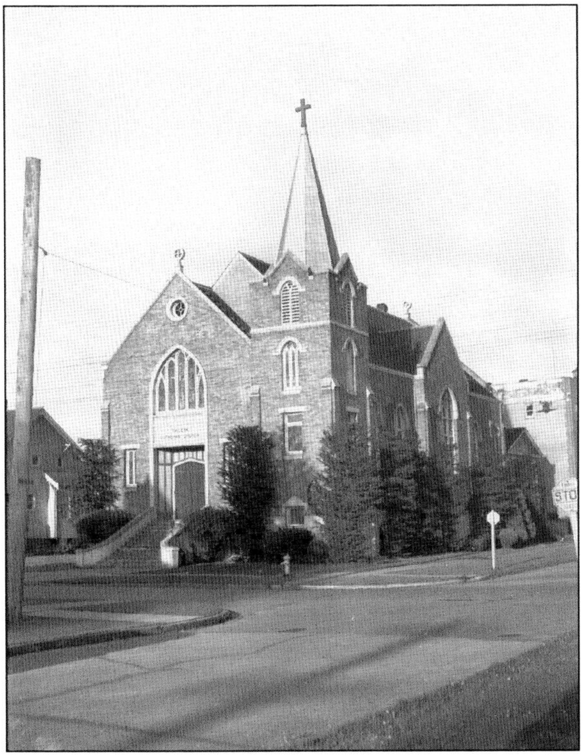

Salem Evangelical Lutheran Church was founded in 1913. All services were in Swedish until 1917, when they began offering one English-language service on the last Sunday of each month. In 1918, Salem purchased property at Third and Snoqualmie Streets and began construction of a church. The sanctuary, photographed around 1940, was completed in 1923. Pastor Martin L. Swanson, seen below (far right) in 1924, was the pastor for Salem from 1922 until 1929. The last regular Swedish-language service was held in 1936. In 1997, the congregation moved to a new church at the corner of Hoag and Laventure Roads. In 1998, the building on Third Street became the home of the Gathering Christian Church, who rented the church before purchasing it in February 2000. (Left, photograph by Bill Forman, courtesy of Mark Iverson; below, courtesy of John Kamb Jr..)

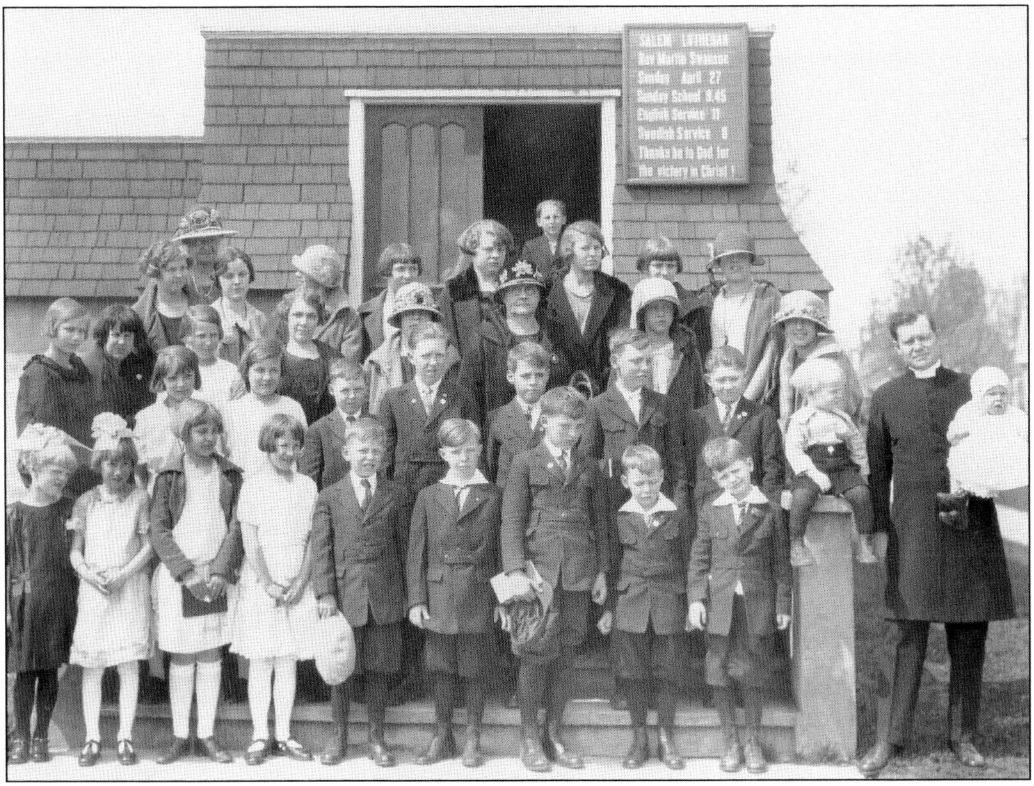

Under the leadership of Rev. John E. Noftsinger, the church undertook sponsorship of a Boy Scout Troop. The troop had begun in 1913, during the very early years of the scouting movement, and had a difficult time finding a local sponsor. To this day, the church still sponsors Boy Scout Troop No. 1, the oldest continually active troop in the State of Washington. (Courtesy of Mount Vernon First Baptist Church.)

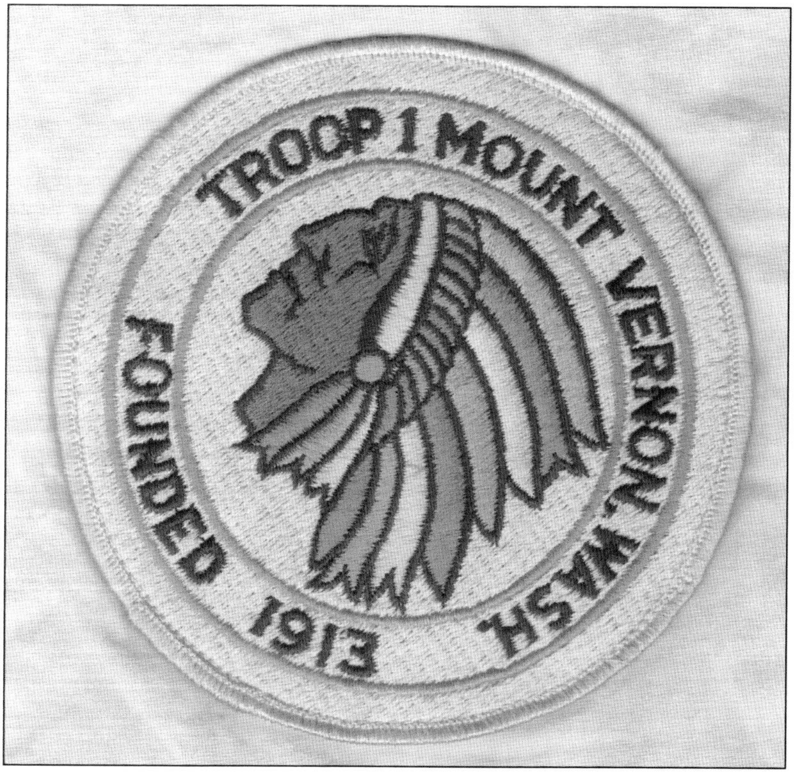

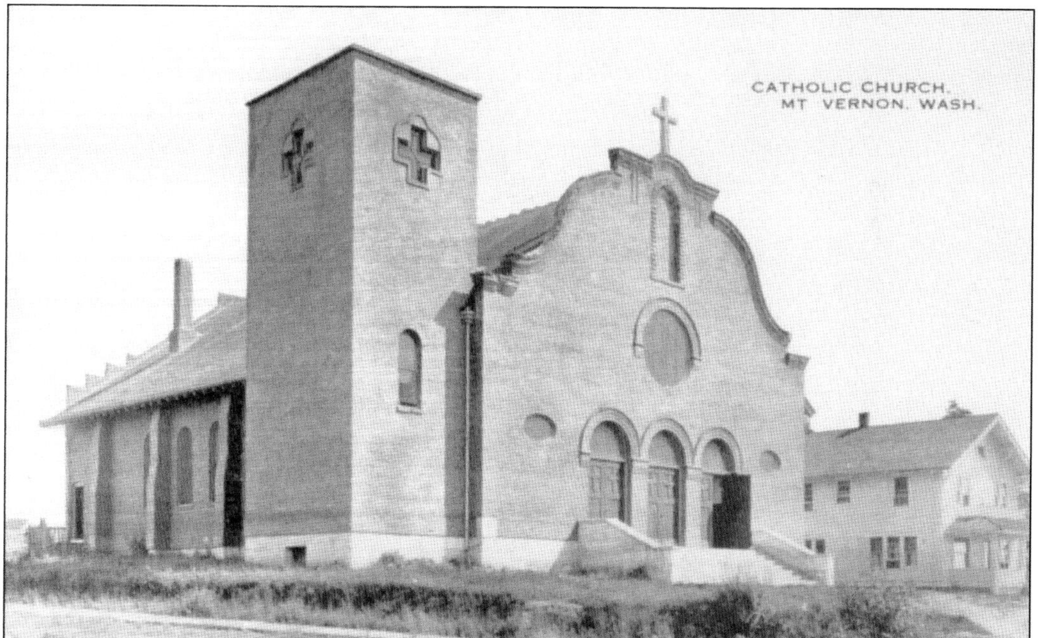

In 1899, Immaculate Conception officially began as a mission church in Mount Vernon. Catholics were brought to the area earlier due to their extensive missionary work with Indians. Immaculate Conception's brick church was built on Fifth Street in 1913, "on the brow of the hill overlooking the city." It replaced an earlier wooden Catholic church. (Courtesy of John Kamb Jr..)

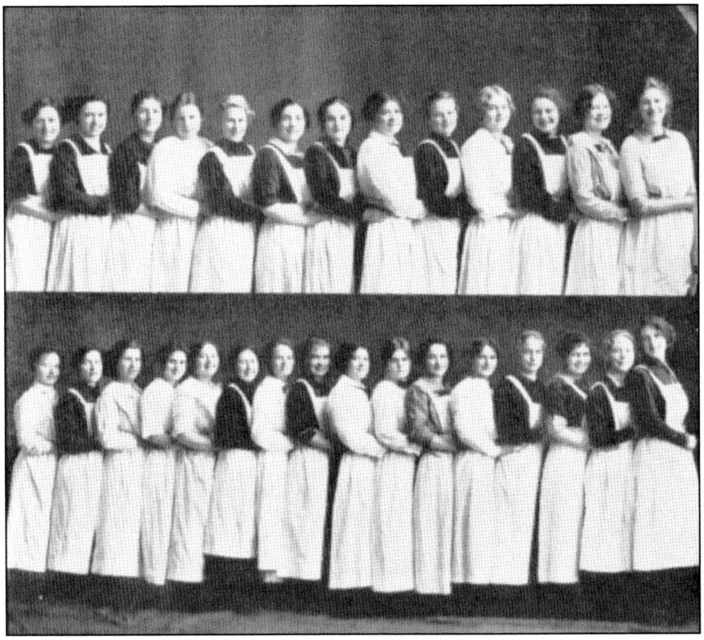

Pictured at left are unidentified members of the Mount Vernon High School 1913 home economics class. Taught by Ida H. Wexler, it was the first home economics class at the school. A write-up in the *Skagina* yearbook explains, "Home economics is divided into two separate subjects—Domestic Science and Domestic Art, the former commonly called 'cookin' and the latter 'sewin'." (Courtesy of the Mount Vernon School District.)

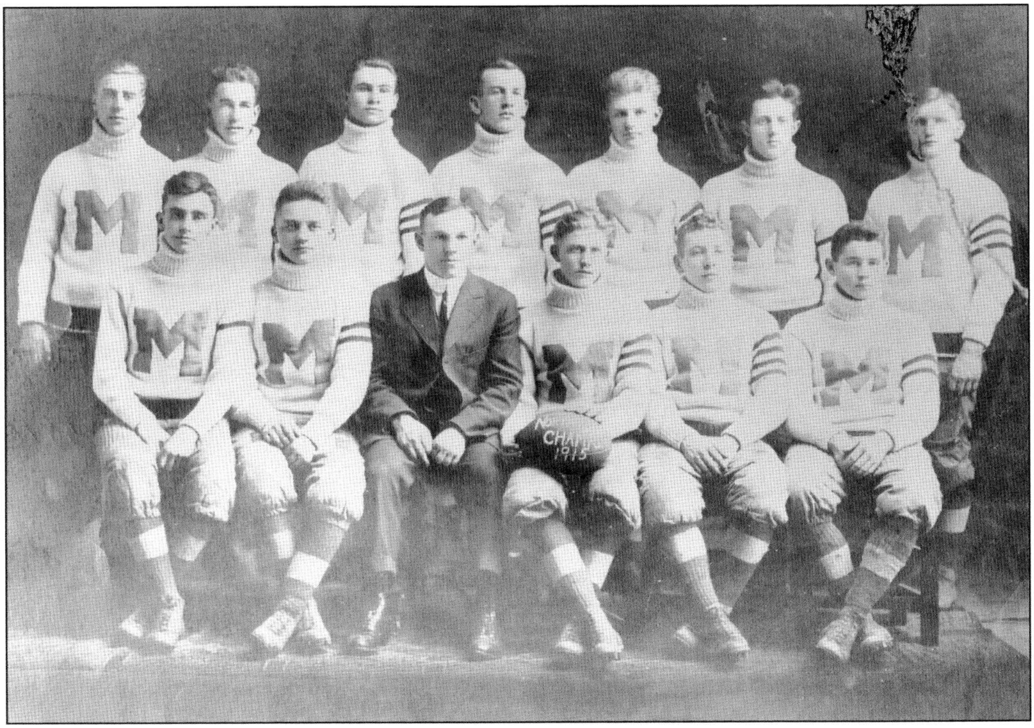

Pictured here is the Northwest Champion Mount Vernon High School football team of 1915. The members include, from left to right, (first row) Roland M. Crow (right end), Fred L. Oars (quarterback), Frank H. Jenne (coach), Roy O. Starr (center), ? Kline (left halfback), and Harry Tait (left end); (second row) Will Gilkey (left halfback), Ernest Carlyle White (right end), Martin Lockhart (fullback), George Raymond Pedersen (right guard), Carl Schroeder (right halfback), Frank Storie (left tackle), and Frank Ewing (right tackle). (Courtesy of the Roger Fox collection.)

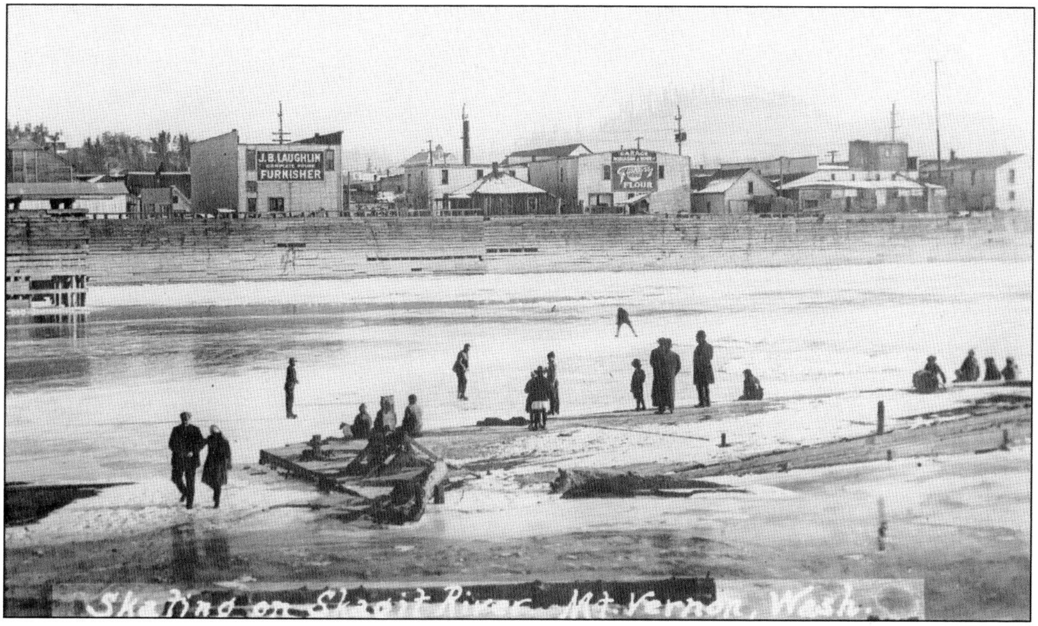

In February 1916, there was a record freeze and snowfall. Not only did the Skagit River freeze over at Mount Vernon, the saltwater where the Skagit meets Puget Sound also froze. Adventurous citizens, seen above, took advantage of the rare river freeze to ice-skate. In the background is Mount Vernon's waterfront. Seen below is a freshly cleared downtown sidewalk, outside the office of Dr. Harry Thornton D'Arc. A native of Canada, Dr. D'Arc arrived in Mount Vernon in January 1915 and was appointed the city physician that same year. (Above, courtesy of the Roger Fox collection; below, courtesy of John Kamb Jr..)

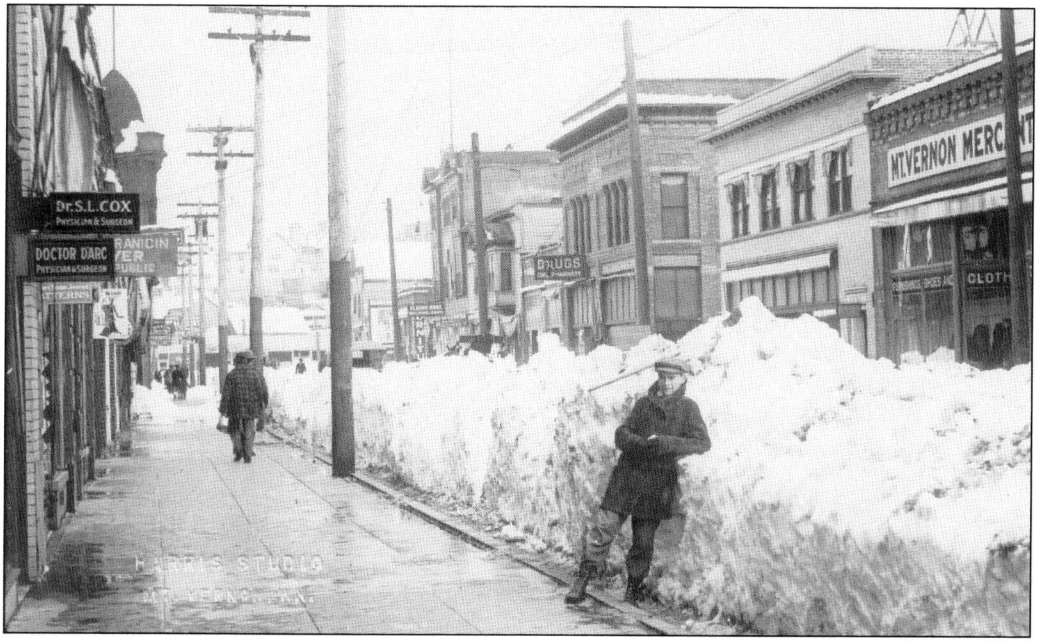

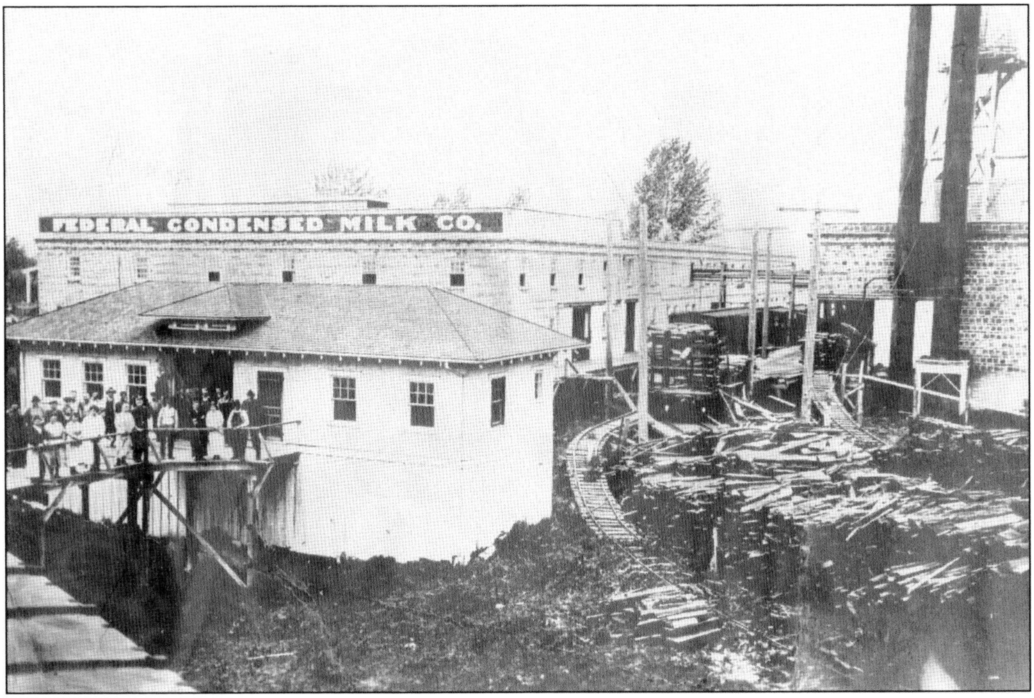

In April 1917, the Federal Condensed Milk Company was built south of downtown at the corner of South Main and Milwaukee Streets. Unable to compete with the near monopoly of the Carnation Company, Federal Condensed Milk Company went out of business in 1919, dropping its 124 suppliers. The Skagit County Dairymen's Association bought a creamery in Burlington to take the milk of the members that had been dropped by Federal. Each year, members paid $10 per cow to belong to the cooperative. The Burlington plant became the association's headquarters, and in April 1922, the group bought the defunct Federal plant as their second local facility for the production of their Darigold brand. Seen below in 1946, the association remained in Mount Vernon for many years. (Above, courtesy of the Roger Fox collection; below, photograph by Bill Forman, courtesy of Mark Iverson.)

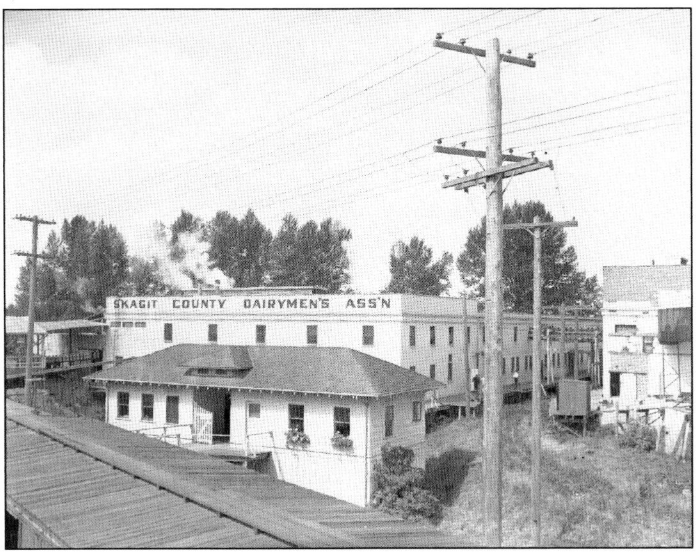

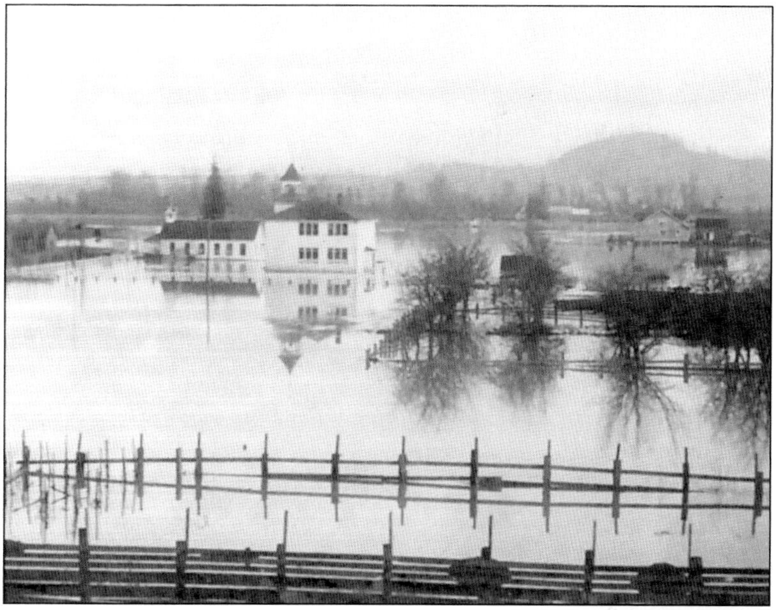

The Harmony Baptist Church (the white structure to the left, constructed in 1901) and the two-story New Harmony School (built in 1908) are surrounded by floodwaters in this 1917 image. Little Mountain is seen in the background. This flood was estimated to have caused approximately $500,000 worth of damage to the area. (Courtesy of John Kamb Jr..)

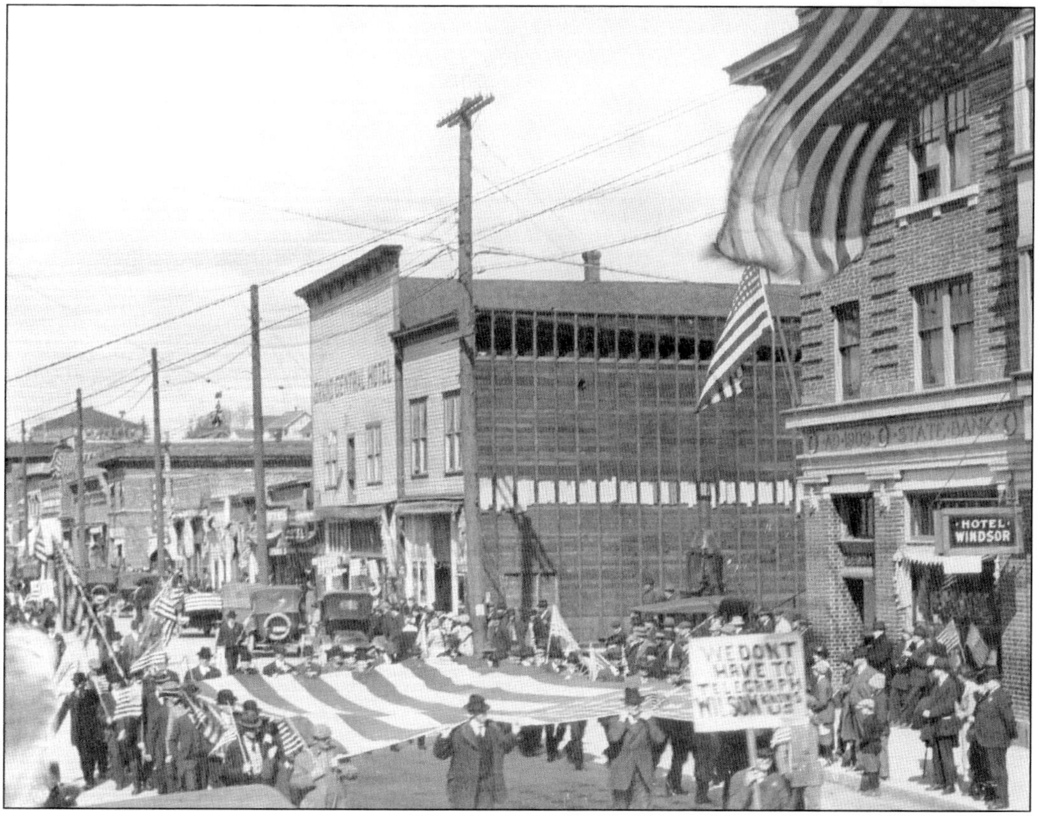

On April 6, 1917, the United States entered World War I. On April 11, Mount Vernon hosted a Flag Parade and Public Pledge of Loyalty, seen here. It is believed that the large flag being carried is the original Pioneer Flag, which was used in patriotic community celebrations until its final parade on Armistice Day, November 11, 1924. (Courtesy of the Roger Fox collection.)

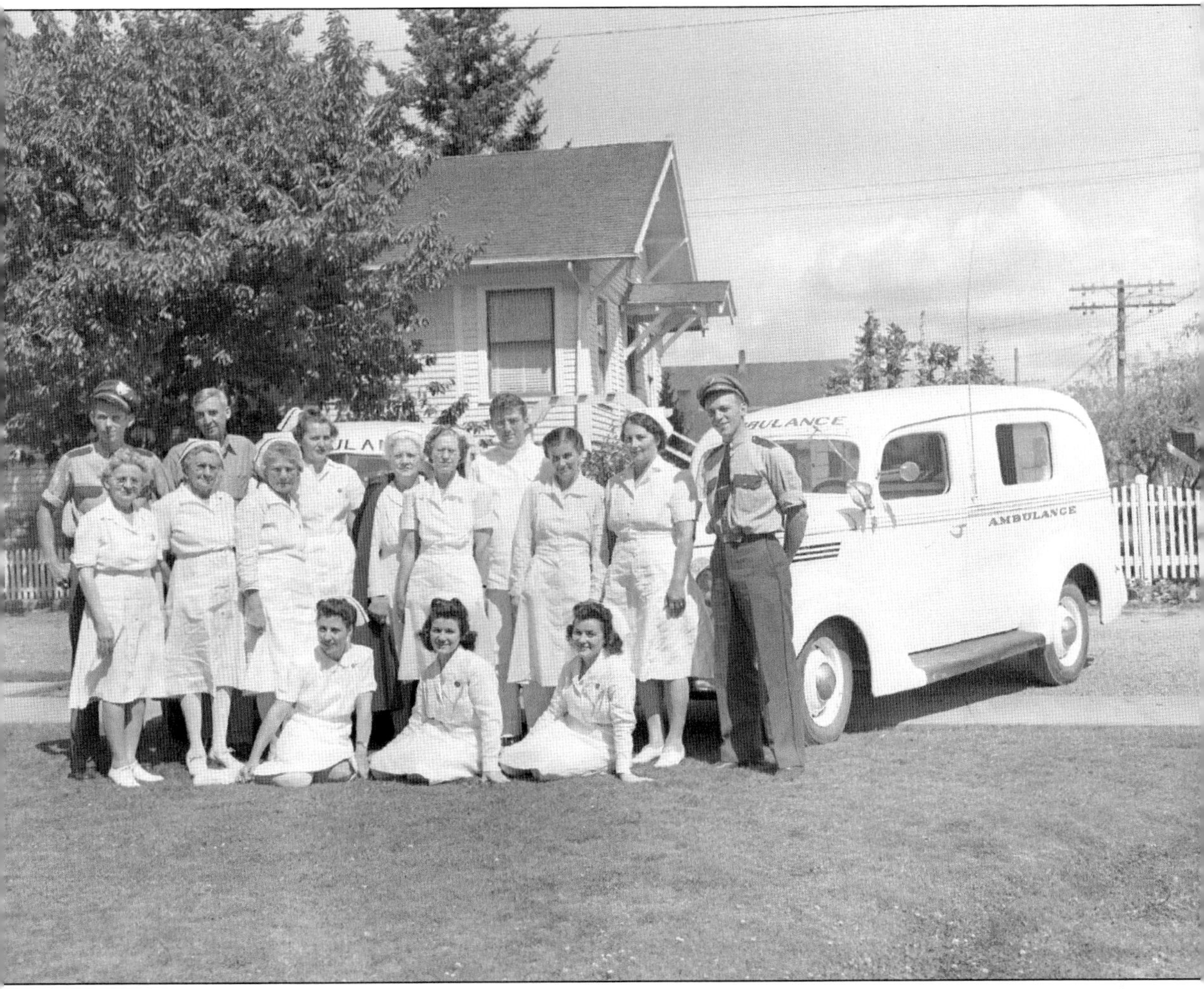

On June 9, 1919, Mount Vernon General Hospital, founded by Dr. Harry Thornton D'Arc, opened. Previously, the closest hospital had been in Burlington. Dr. D'Arc ran the hospital for several years before selling it. At one point, it was owned and operated by a local group of doctors. In 1946, it became a nonprofit community hospital and was renamed Skagit General Hospital. In 1958, the facility had 37 beds and 12 bassinets. The hospital building, at 307 East Division Street, survives today as an apartment complex. Meanwhile, in 1922, Harriet Wade Rowley (born October 12, 1880) opened the Mount Vernon Maternity Hospital at 1011 East Division Street. In 1934, she added the first surgery unit, and the hospital eventually grew into Rowley General Hospital, with 54 beds and 12 bassinets. Pictured here are unidentified members of the Rowley Hospital staff in 1946. Skagit General closed in 1958 with the opening of Skagit Valley Hospital, and Rowley closed shortly thereafter. (Photograph by Bill Forman; courtesy of Mark Iverson.)

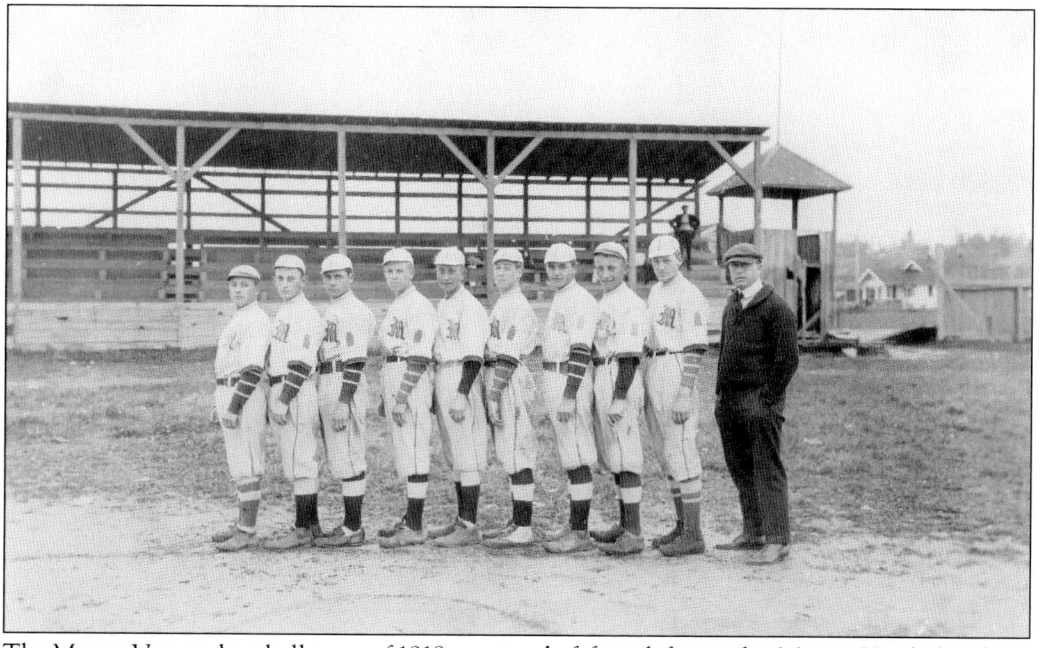

The Mount Vernon baseball team of 1918 consisted of, from left to right, Martin Herrle (pitcher), Harold Good (captain, catcher), Emerson Wolf (center field), Richard Welts (shortstop), Gordon Good (second base), Richmond Valentine (left field), Jerome Hannaford (first base), Howard Good (third base), Lewis Crogstad (right field), and W.H. Talley (coach). The team's record was 4-2, beating Burlington (twice), Bellingham, and Anacortes but not Sedro-Woolley. (Courtesy of John Kamb Jr..)

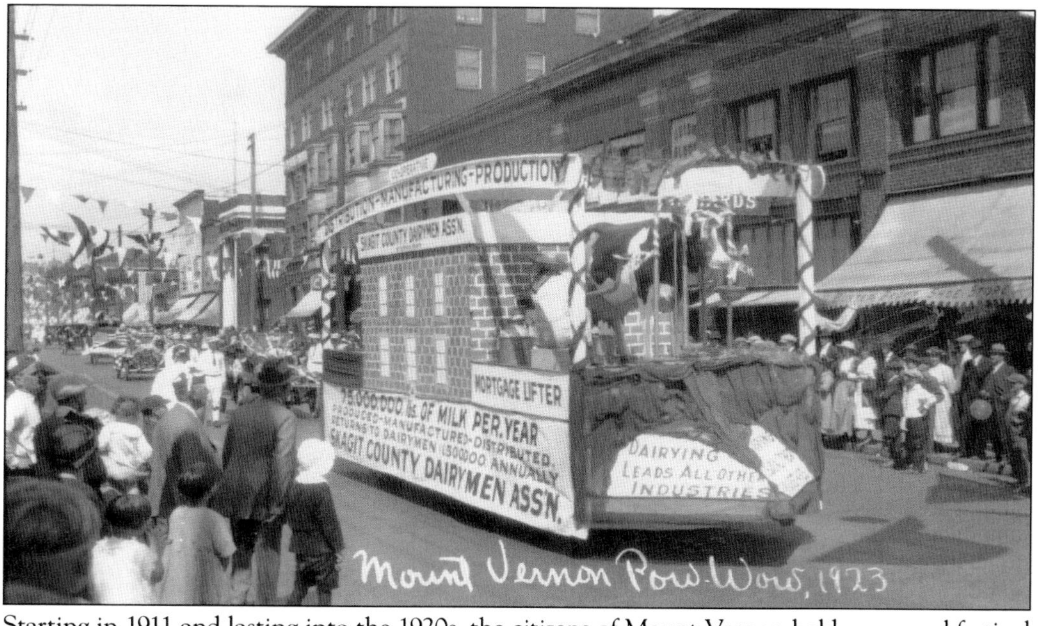

Starting in 1911 and lasting into the 1920s, the citizens of Mount Vernon held an annual festival called the Mount Vernon Pow-Wow. The festival seems to have been an exercise in civic pride along with an excuse for fun. In the 1930s, the Pow-Wow festival became Hickory Hat Days and continued into the 1950s. Seen here is a Pow-Wow parade float from 1923. (Courtesy of John Kamb Jr..)

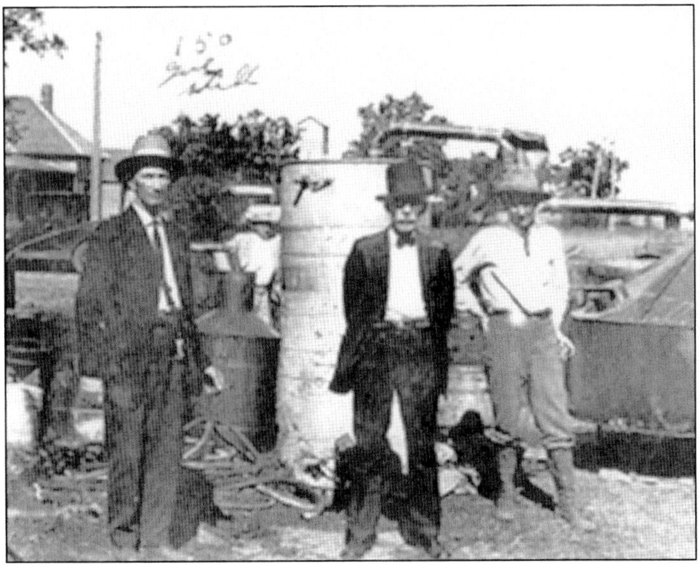

On November 3, 1914, Washington voters approved Initiative Measure No. Three, prohibiting the manufacture and sale (although not the consumption) of liquor statewide. Washington was among 33 of the existing 48 states that adopted Prohibition laws at the state level before the 18th amendment was ratified in 1919. Illegal moonshine stills, like the one seen here, were targets of local law enforcement. (Courtesy of the Skagit County Auditor's Office.)

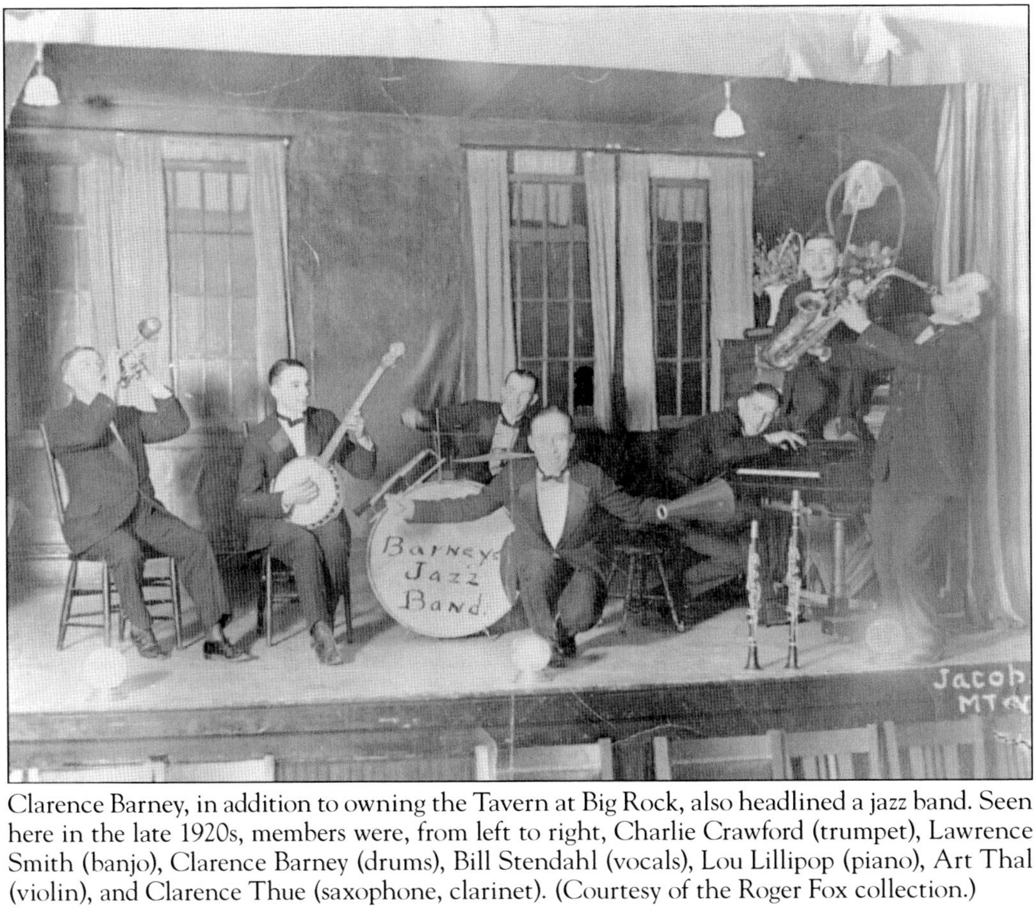

Clarence Barney, in addition to owning the Tavern at Big Rock, also headlined a jazz band. Seen here in the late 1920s, members were, from left to right, Charlie Crawford (trumpet), Lawrence Smith (banjo), Clarence Barney (drums), Bill Stendahl (vocals), Lou Lillipop (piano), Art Thal (violin), and Clarence Thue (saxophone, clarinet). (Courtesy of the Roger Fox collection.)

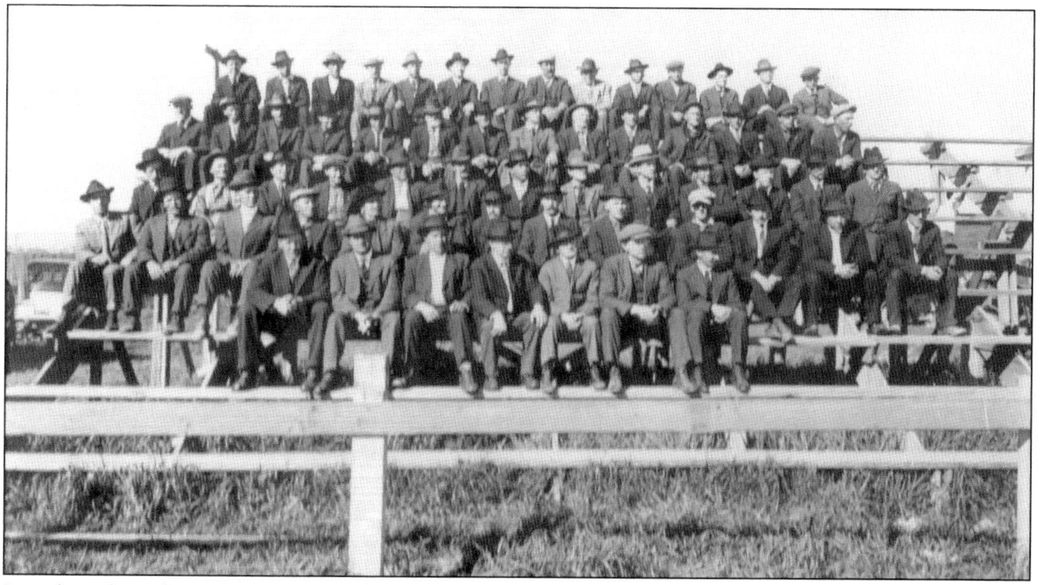

Local seed producers are not as well known as their tulip-growing counterparts, but in the 1950s, the area was the largest cabbage-seed producer in the world. Particularly after World War II, seed producers played a significant role in international agriculture. Cabbage, beet, rutabaga, carrot, and turnip seeds were all cultivated here. This image shows the Seed Growers Association in the early 1920s, around the time of its founding. (Courtesy of the Basye family.)

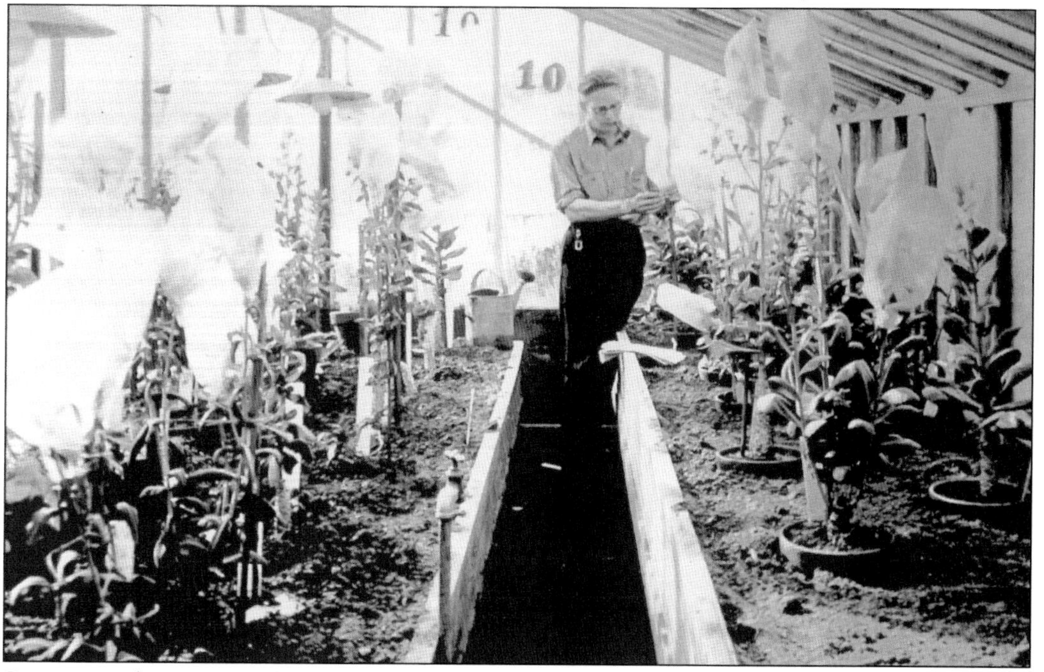

Starting in the 1920s, makeshift greenhouses, like the one seen here, were set up by Washington State University at the fairgrounds in Mount Vernon for research and community education. As part of the effort to help rebuild agriculture in war-torn Europe, funding was approved to build the lab that became the Western Washington Research and Extension Unit in west Mount Vernon in 1947. (Courtesy of the WSU Skagit County Extension.)

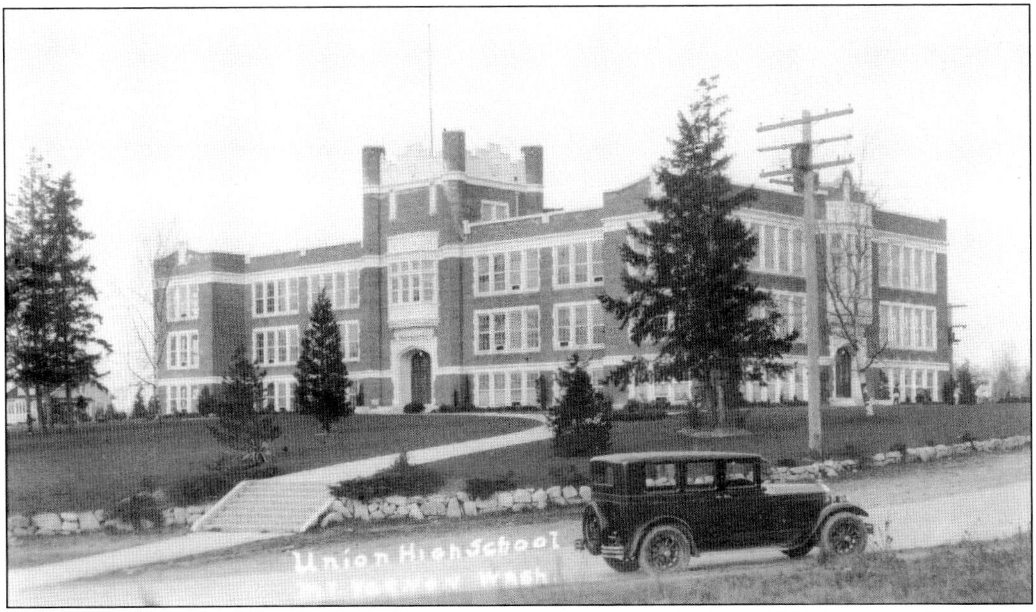

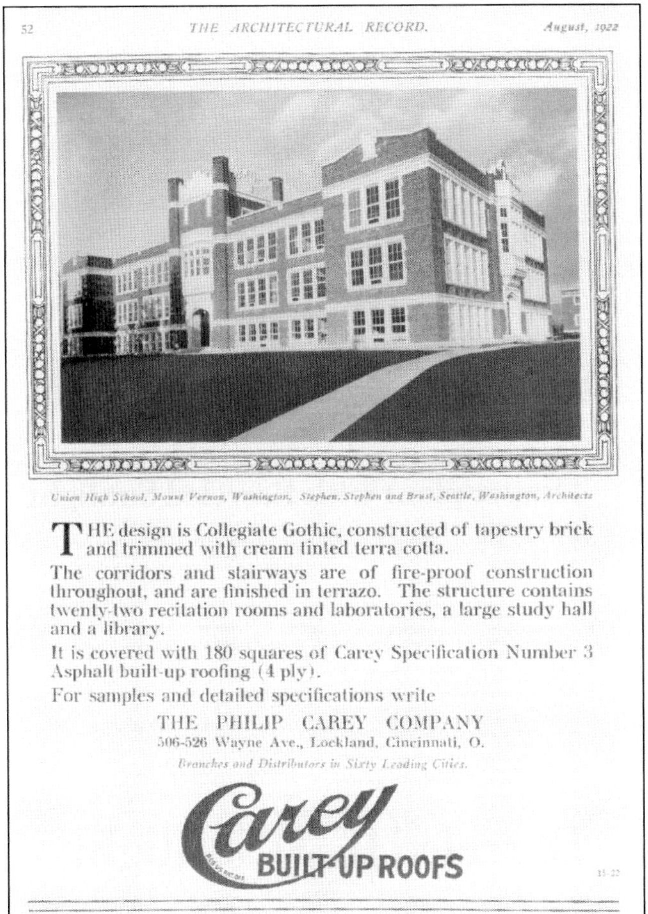
52 THE ARCHITECTURAL RECORD. August, 1922

Union High School, Mount Vernon, Washington. Stephen, Stephen and Brust, Seattle, Washington, Architects

THE design is Collegiate Gothic, constructed of tapestry brick and trimmed with cream tinted terra cotta.

The corridors and stairways are of fire-proof construction throughout, and are finished in terrazo. The structure contains twenty-two recitation rooms and laboratories, a large study hall and a library.

It is covered with 180 squares of Carey Specification Number 3 Asphalt built-up roofing (4 ply).

For samples and detailed specifications write

THE PHILIP CAREY COMPANY
506-526 Wayne Ave., Lockland, Cincinnati, O.

Branches and Distributors in Sixty Leading Cities.

Carey BUILT-UP ROOFS

15-22

In June 1921, Mayor B.D. Moody requested that all businesses close for two hours to celebrate the laying of the cornerstone of the new Union High School. The program began with a parade through downtown to the school site, followed by dedication ceremonies, and concluded with a baseball game and dance in the evening. The school, completed in 1922, served 14 different school districts. Union was considered a model rural high school. In 1924, pupils from farm homes numbered 242 and those from nonfarm homes numbered 170. It was said that Union High School District One was "one of the largest and richest districts in the Northwest." The building itself was a grand accomplishment, as its use in this advertisement attests. (Above, courtesy of John Kamb Jr.; left, courtesy of the Mount Vernon Mayor's Office.)

In the early 1920s, seven trucks operating out of Mount Vernon served as transportation to the high school from the rural districts. Over 200 students were transported this way, and routes varied from 9 to 35 miles, round-trip. In 1924, the district's transportation costs were $9,500. This 1925 image captures a truck on the 24-mile Milltown route. (Courtesy of Ed Marlow.)

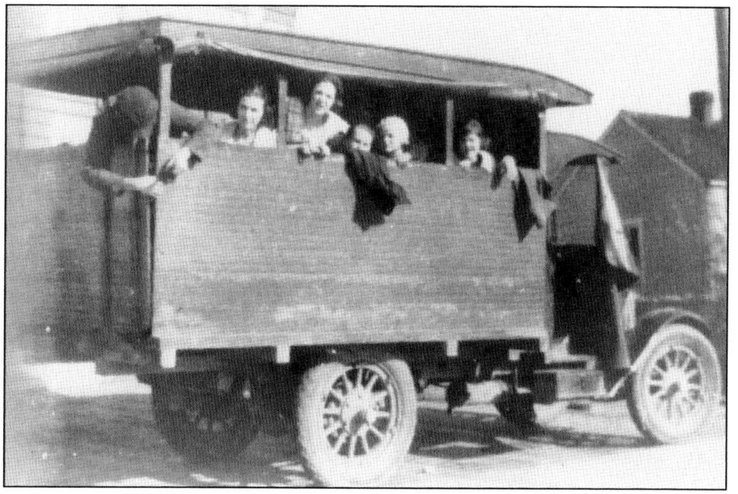

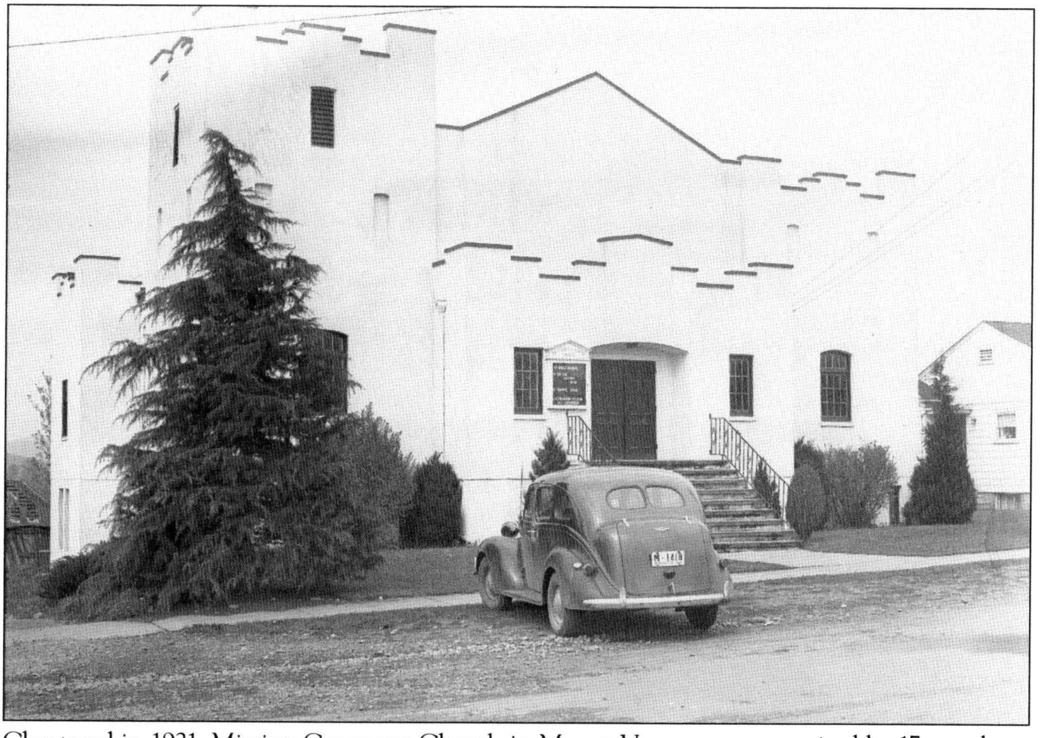

Chartered in 1921, Mission Covenant Church in Mount Vernon was organized by 17 members. In January 1928, its name was changed to Bethany Mission Church. Seen here in the early 1940s, it was built on Evergreen Street, with construction starting in 1922 and finishing in 1937. The year of its completion, the name was changed one final time to Bethany Covenant Church. (Photograph by Bill Forman; courtesy of Mark Iverson.)

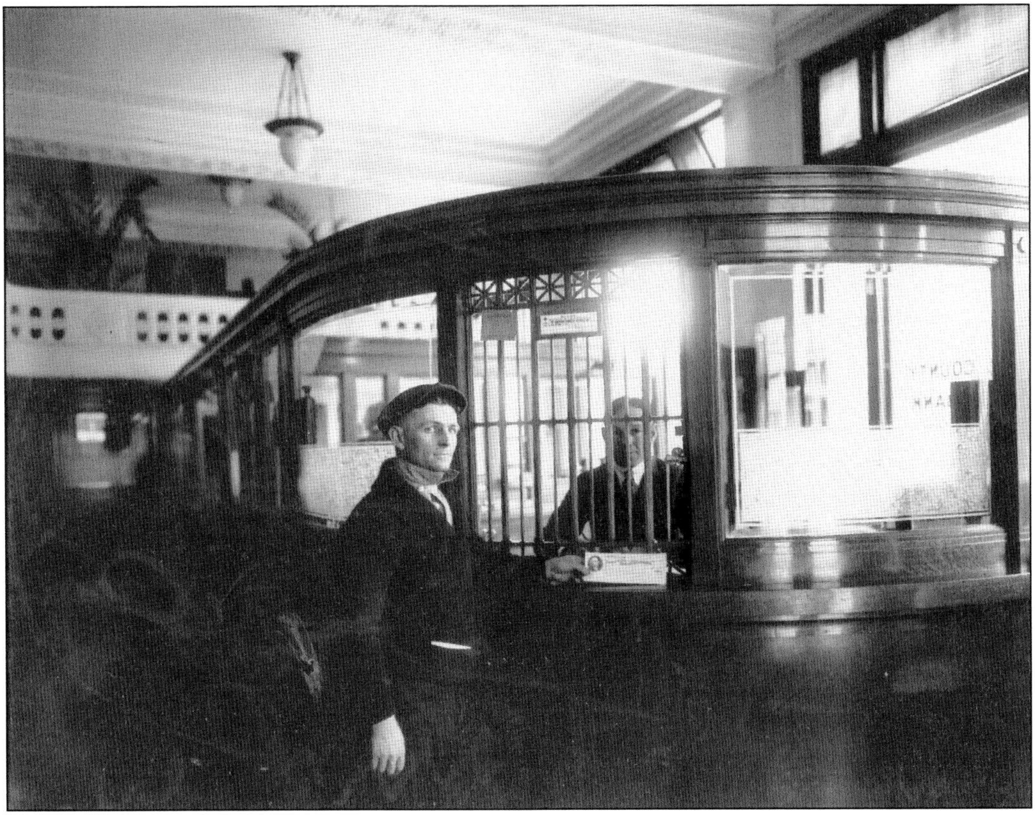

The 1920s photograph above shows Carl Noce (left) and Frank Jenne (right) inside Skagit County National Bank. Organized in 1911 as the Skagit County Savings & Trust Company, it engaged mostly in loans and investments until 1918, when it activated its banking charter. In 1922, a new Neoclassical bank, seen below, was erected on the corner of First and Myrtle Streets at the former site of the Miller Hotel. In May 1929, the bank merged with Mount Vernon National Bank (located directly across Myrtle Street at the corner of the Windsor Hotel). The new consolidated bank was known as the Skagit National Bank of Mount Vernon. In 1934, it was absorbed as a branch of Seattle First National Bank, which operated at this location until 1970. (Both, courtesy of the Roger Fox collection.)

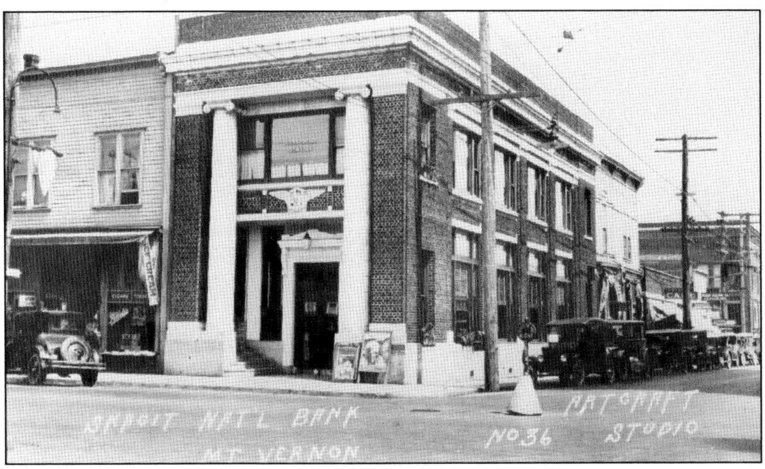

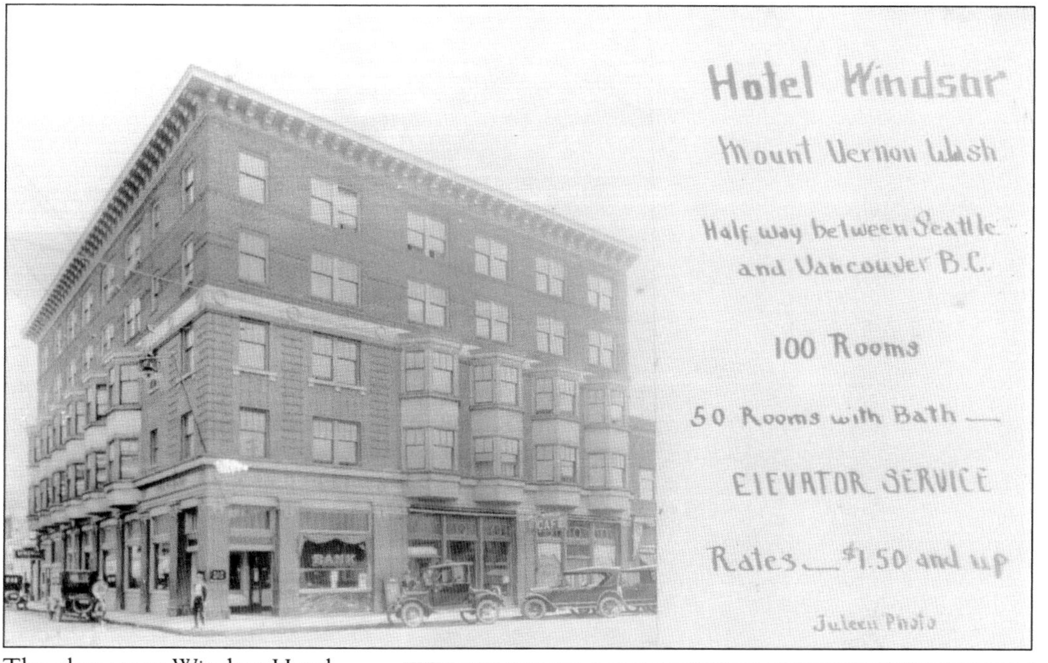

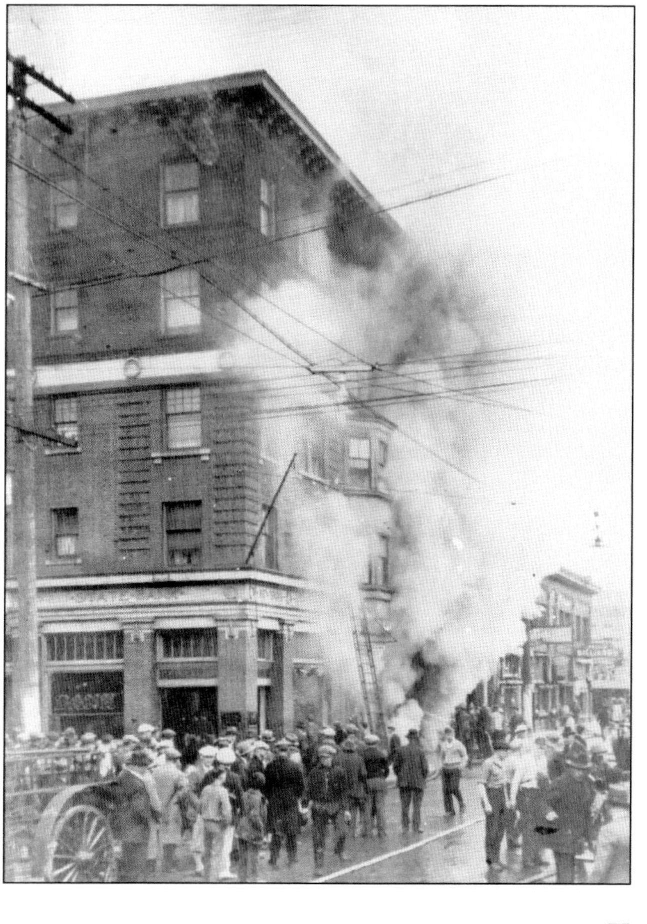

The glamorous Windsor Hotel, at the corner of Myrtle and First Streets, was built in 1909. Guests came by way of stern-wheeler, train, and automobile. Originally, the hotel was located on two upper stories, with commercial property on the ground floor—most notably, the Mount Vernon State Bank (established in 1906), located on the northwest corner. As the Windsor's popularity increased, the hotel expanded by adding two more floors, as seen on this postcard, in 1922. Unfortunately, on the morning of March 23, 1926, a devastating fire destroyed most of the building. Amazingly, the bank suffered no damage and reopened rather quickly. A replacement hotel opened on January 8, 1927, and was renamed the President Hotel. (Above, courtesy of John Kamb Jr.; right, courtesy of Ed Marlow.)

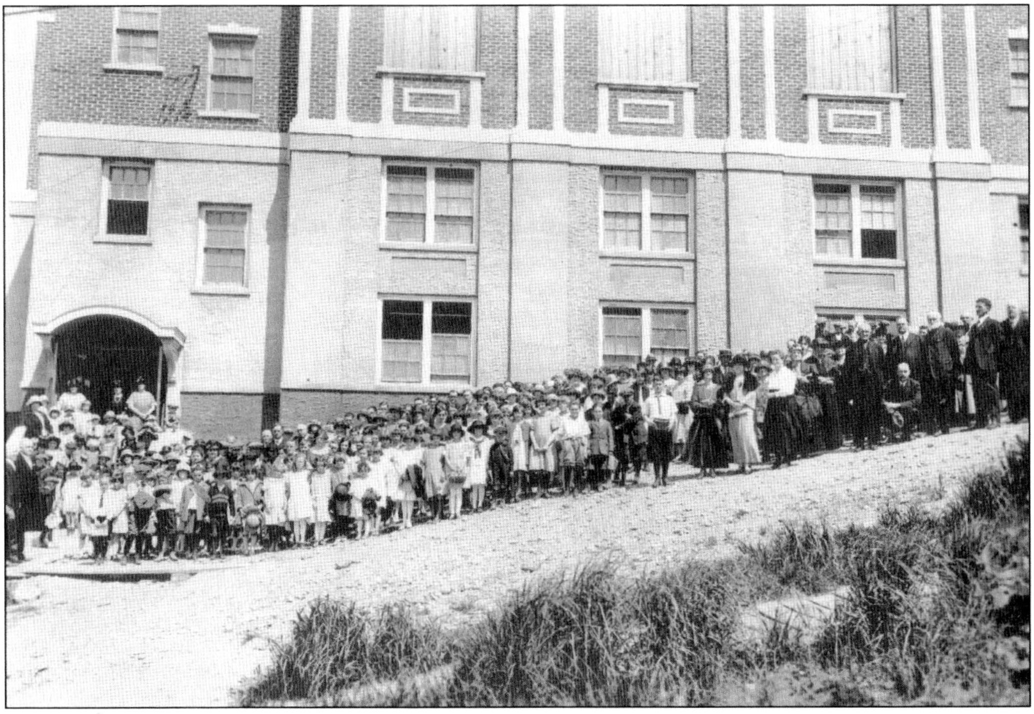

Above, the congregation of Mount Vernon First Baptist Church stands before the south wall of their new church on Easter 1924. This was the third and final location for the church. Note that Division Street is unpaved, and the upper windows are covered in boards. In 1923, it became necessary to stop work on the building because of a lack of funds. The upper sanctuary was left unfinished, while the lower auditorium was completed. The congregation moved into the lower level of the new church in January 1924, but the upper sanctuary was not completed until 1935. Below, a service is held in the lower auditorium. (Both, courtesy of Mount Vernon First Baptist Church.)

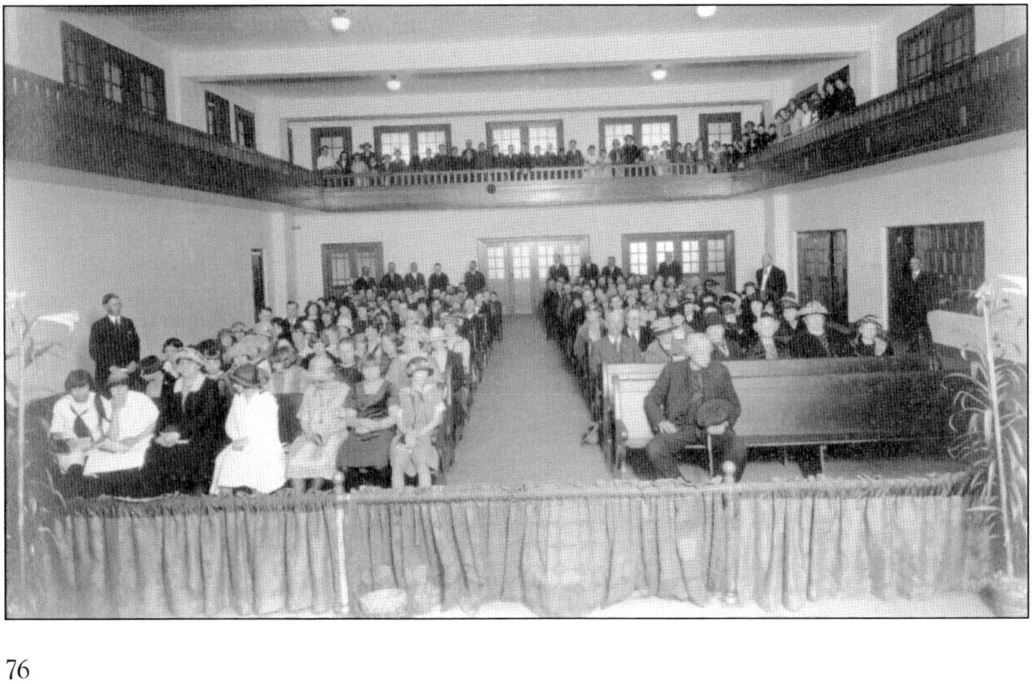

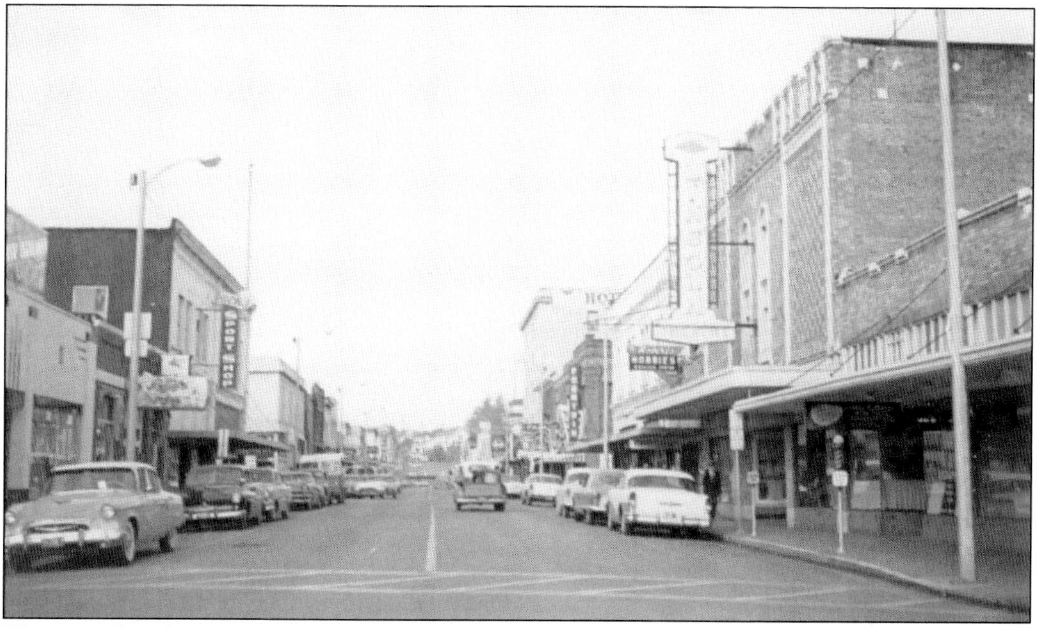

The Lincoln Theatre, visible on the right side of the street in the 1960s image above, was built in 1926. Quite unusual for the Northwest, the Lincoln, known as a period theater, had a Spanish motif. The manager, Edwin Halberg, ignored the pleas of his friends to follow convention and make it Egyptian. The original building cost $100,000, the organ cost $22,500, the furnishings and equipment cost $32,500, and the 1926 payroll was $160 a week. The nonprofit Lincoln Theatre Center Foundation was created in 1987 to support the theater. Project 85, a fundraising effort to replace the marquee in the style of the original, was launched in 2011 (in celebration of the Lincoln's 85th anniversary). The new sign was installed on May 30, 2012. (Above, courtesy of John Kamb Jr.; below, courtesy of Jessica Bylund.)

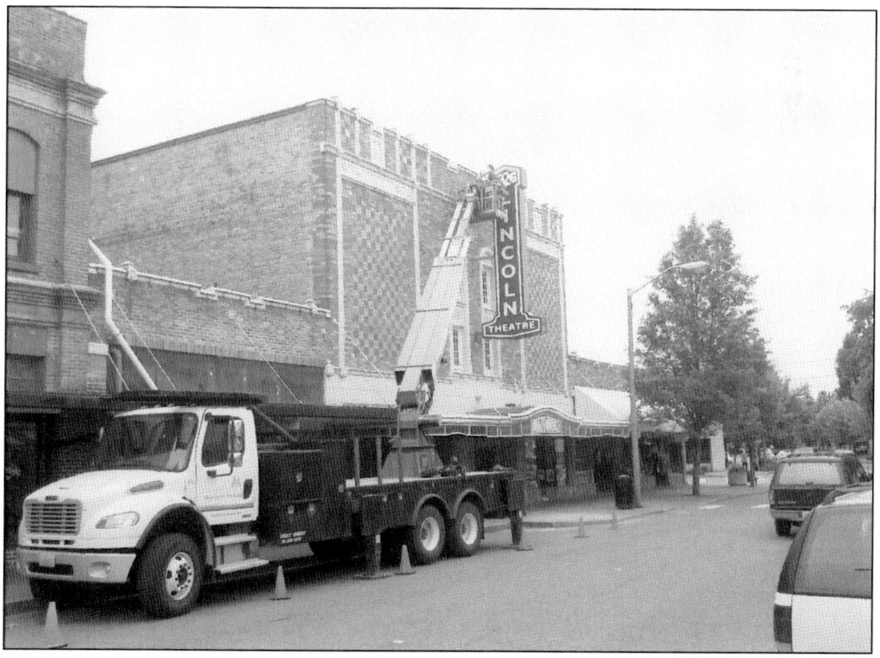

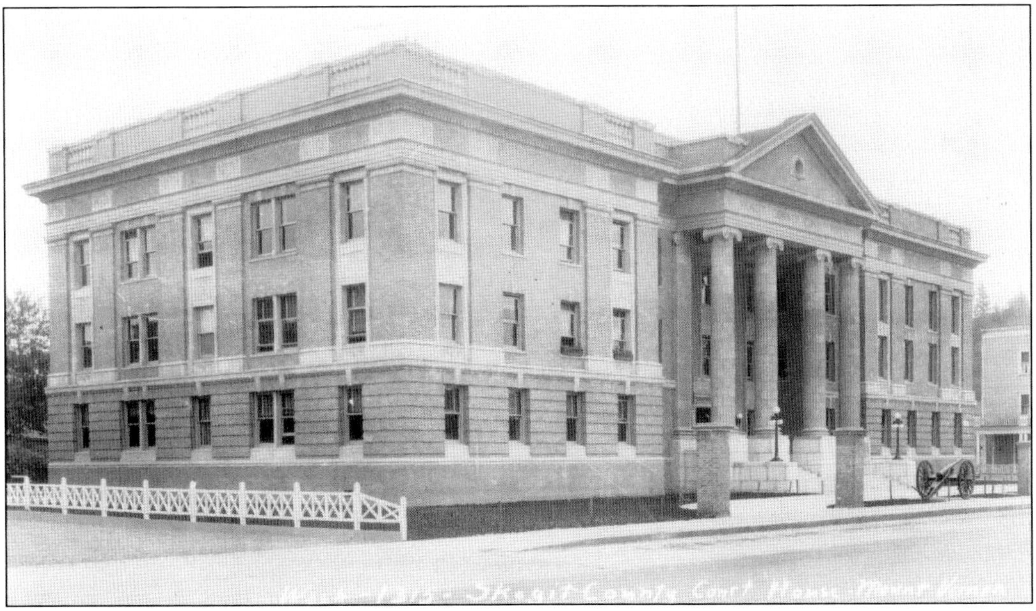

Bellingham architect Thornton F. Doan designed the 45,000-square-foot county courthouse that was built in 1922 on Kincaid Street. On Armistice Day, November 11, 1924, Mount Vernon's 24-by-36-foot American flag, handmade by pioneers in 1877, was put on permanent display in the courthouse lobby through the efforts of the Daughters of the American Revolution. (Courtesy of John Kamb Jr..)

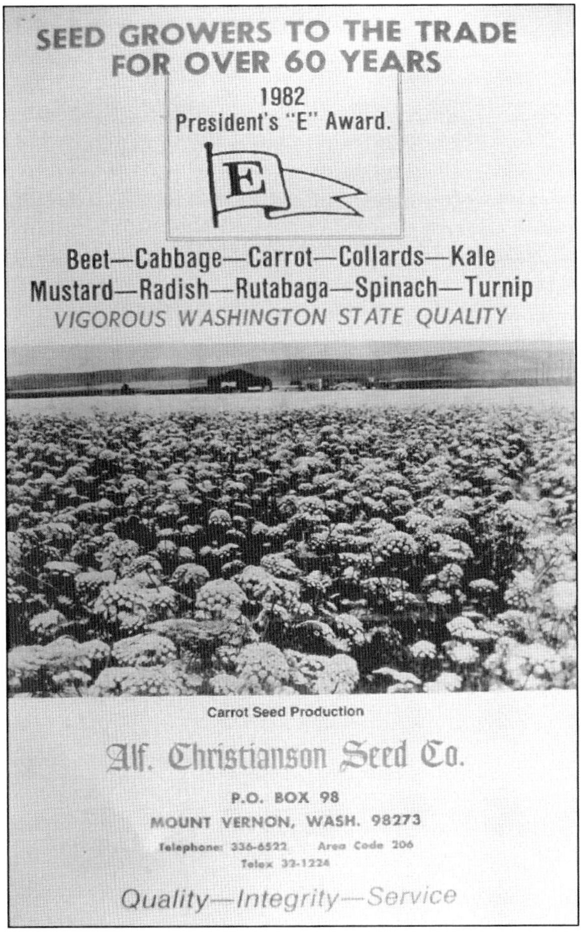

In 1926, after 10 years in Skagit County working for the Seattle-based Charles H. Lilly Company, Alf Christianson established Alf Christianson Seed Company in Mount Vernon. He recognized that the area was ideal for growing cool-weather vegetables like cabbage, beets, turnips, spinach, and carrots. The company became both locally and internationally well known. A Christianson seed packet is seen here. (Courtesy of the WSU Skagit County Extension.)

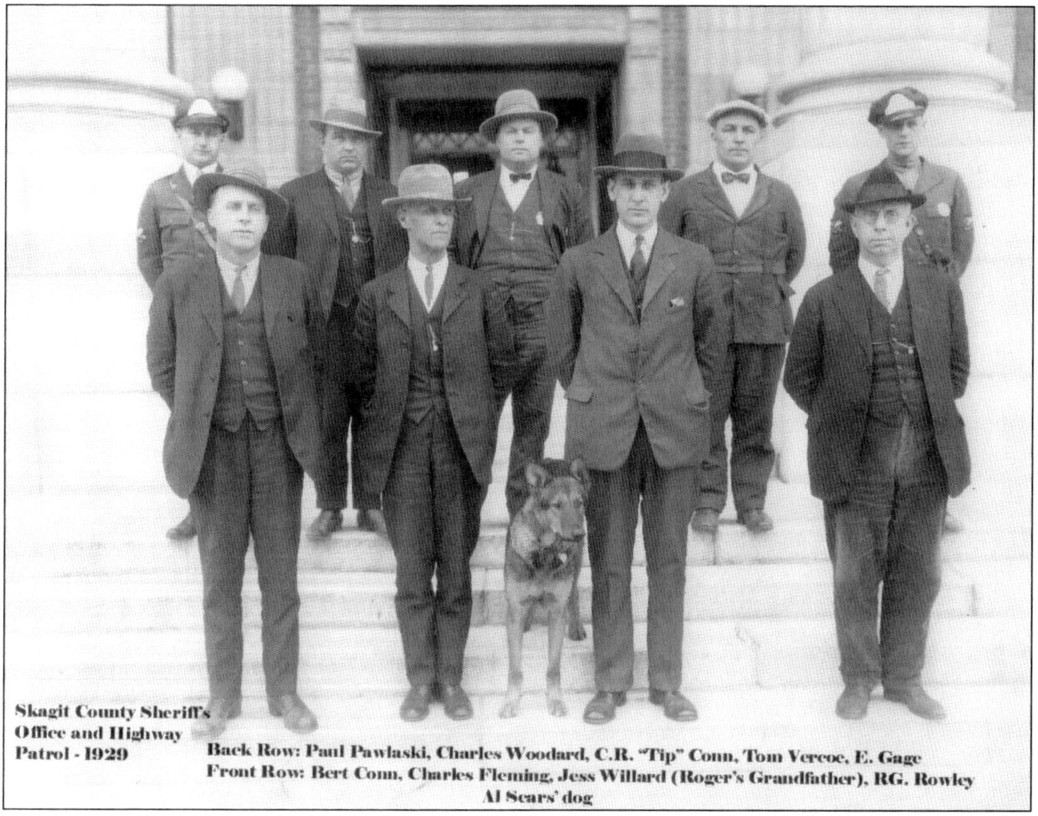

The Skagit County Sheriff's Office and Highway Patrol pose on the steps of the courthouse in 1929. Pictured are, from left to right, (first row) Bert Conn, Charles Fleming, Al Sears's dog, Jess Willard, and R.G. Rowley; (second row) Paul Pawlaski, Charles Woodard, C.R. "Tip" Conn, Tom Vercoe, and E. Gage. (Courtesy of the Skagit County Auditor's Office.)

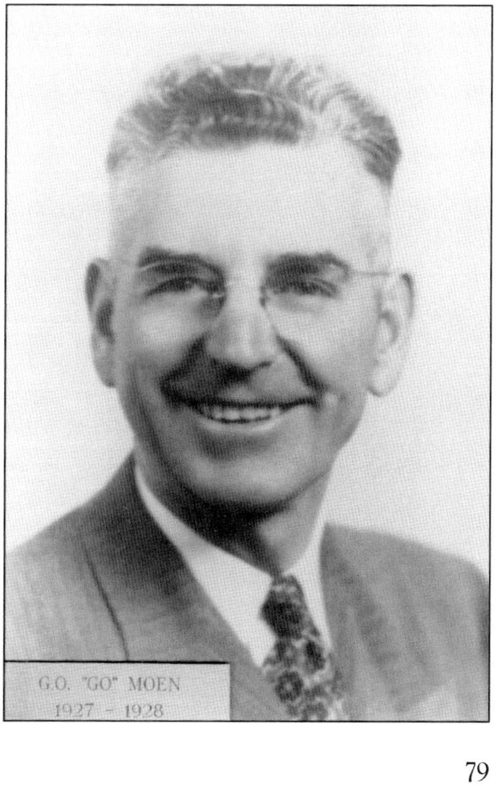

G.O. "Go" Moen was the mayor of Mount Vernon from 1927 until 1928. He founded Mount Vernon Investment, which was located in the President Hotel, and was the Rotary Club president in 1929. Moen came from a family of bankers, and his brothers Guy and Carl were both presidents of different banks in the late 1930s. (Courtesy of the Mount Vernon Mayor's Office.)

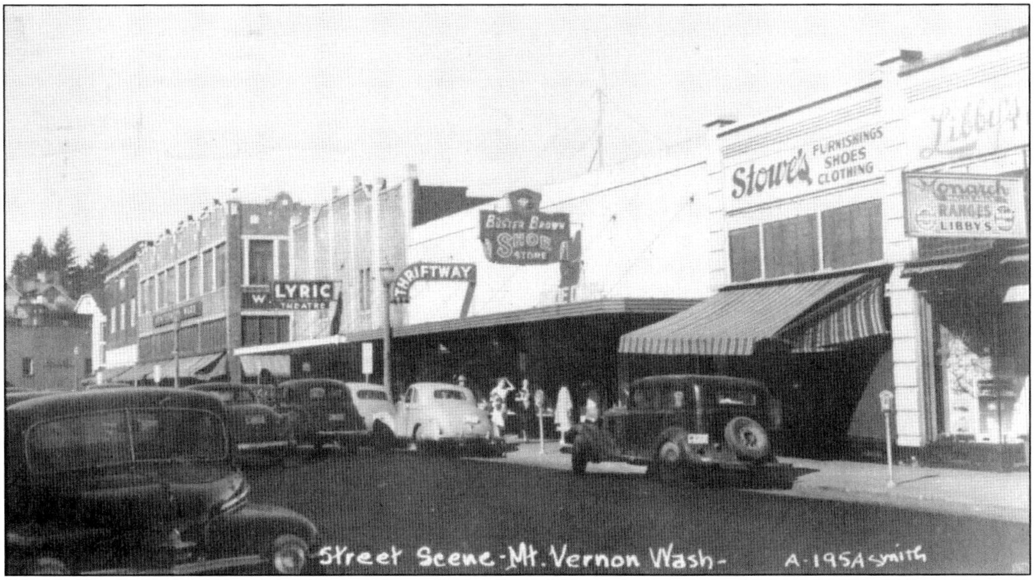

At the corner of First and Washington Streets, where the Opera House had once stood, the Dobson Building was constructed in 1929. The building initially housed a Paramount Theatre, where regular admission was 25¢ for adults and 10¢ for children. Later, the building contained the Roxy Theatre (opened in 1932), the Lyric (seen above, around 1950), and finally, the Lido (opened in 1953). The building operated as a movie house until the early 1990s; it was also used for live events. The photograph below, of a program in 1950, displays the ornate decor and the Paramount *P* (on the left column) from the original theater. (Above, courtesy of John Kamb Jr.; below, photograph by Bill Forman, courtesy of Mark Iverson.)

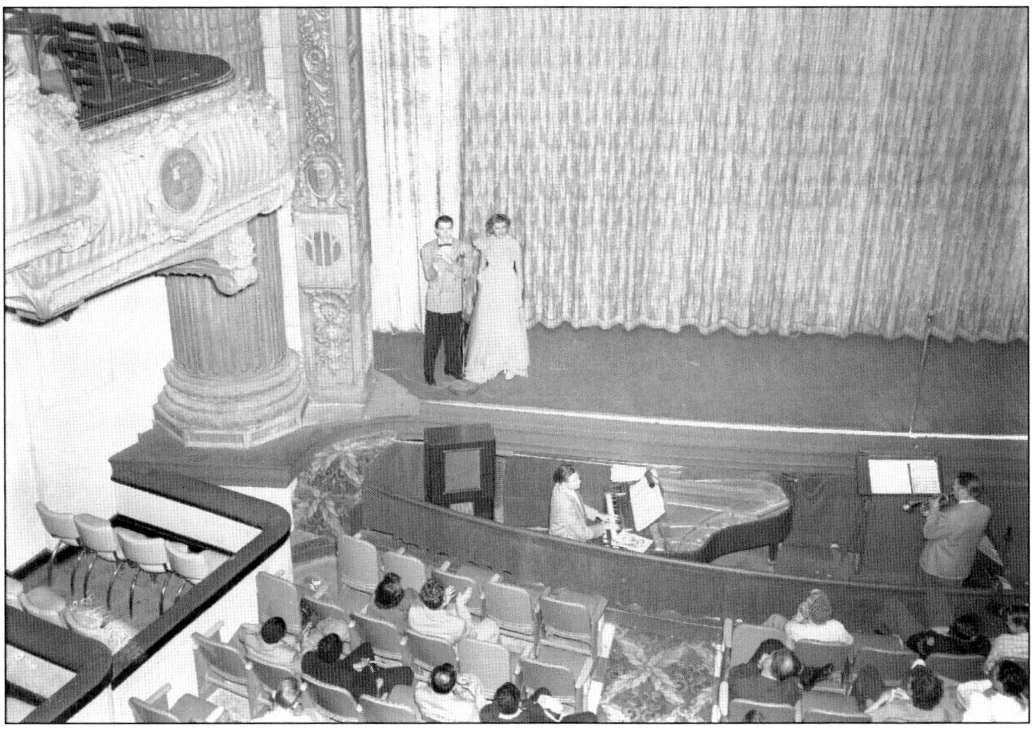

Six

Resilience 1930–1949

The Great Depression made the 1930s a difficult time. Most development in the town was done through federal Works Progress Administration (WPA) projects, including a new post office (1935), a new Lincoln School (1938), a road up Little Mountain (1934), and Hillcrest Park Lodge (1938). There was also some expansion in the school district beyond WPA projects. In 1936, the technical education building was added to the high school campus, and in 1937, Cleveland School was constructed.

It was a turbulent time for the fledgling junior college, which shared the Union High School building. In 1936, enrollment surpassed 100 students, and in the fall of 1939, the college was moved to the newly renovated Roosevelt Elementary building. Unfortunately, enrollment began to decline in 1941 because of the war, and in 1944, the struggling college had to move back to the high school. After the war, the college was awarded $330,000 by the state legislature to fund new facilities, and with the passage of the GI Bill, enrollment began to climb once again.

By 1940, approximately 4,000 farms (with an annual yield of over $9 million) were centered around Mount Vernon. The city was the primary trading area for the fertile valley. Following the Depression, the variety of cash crops in the area, including peas, berries, and potatoes, encouraged the building of several processing plants, which brought much-needed jobs to the community. In 1940, the Moffett Company opened a massive cold-storage plant. A publication from that year foreshadowed one of the area's most prominent crops of today, writing, "Bulb growing is the newest of farm enterprises in the Skagit country. Experiments . . . indicate that the valley is well suited to bulb production."

Despite the hardships of the Depression era, each summer the city sponsored Hickory Hat Days, a multiday community celebration. The first radio station, KBRC, started in 1946 and was known as the "Voice of the Rich Skagit Valley." And the semiprofessional baseball team, the Mount Vernon Milkmaids, could be seen playing at the downtown ballpark that was later named after local World War II hero Sherman Anderson.

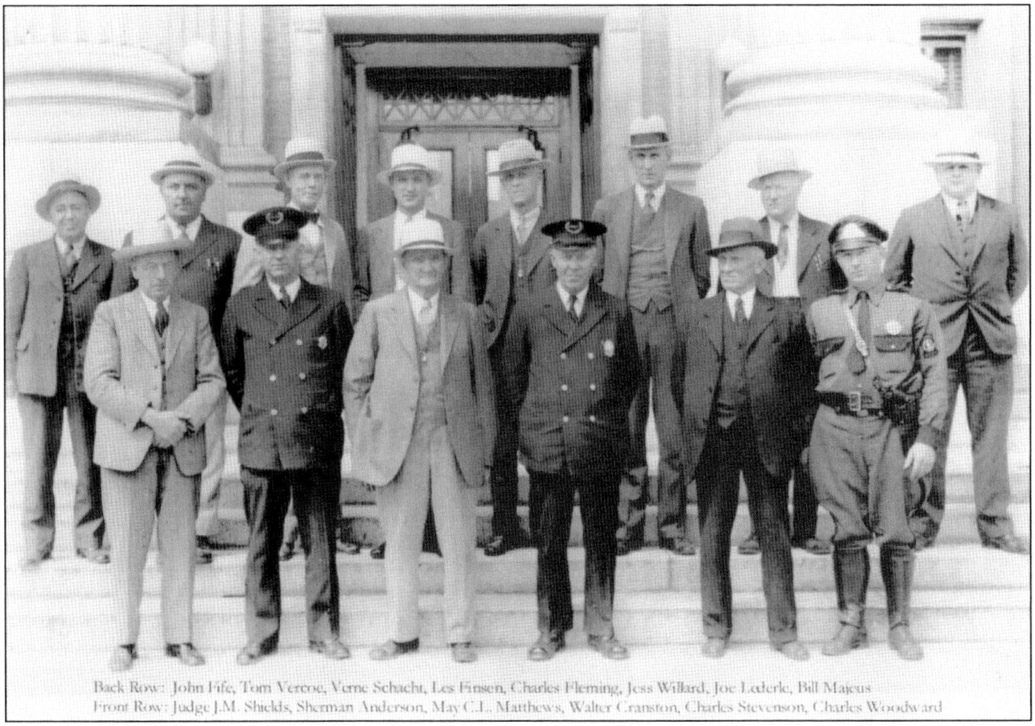

Members of the Skagit County Sheriff's Department and the Mount Vernon Police Department gather in 1931. Pictured are, from left to right, (first row) Judge J.M. Shields, Sherman Anderson, May C.L. Matthews, Chief Walter Cranston, Charles Stevenson, and Charles Woodward; (second row) John Fife, Tom Vercoe, Verne Schacht, Les Finsen, Charles Fleming, Deputy Sheriff Jess Willard, Joe Lederle, and Bill Majeus. (Courtesy of the Skagit County Auditor's Office.)

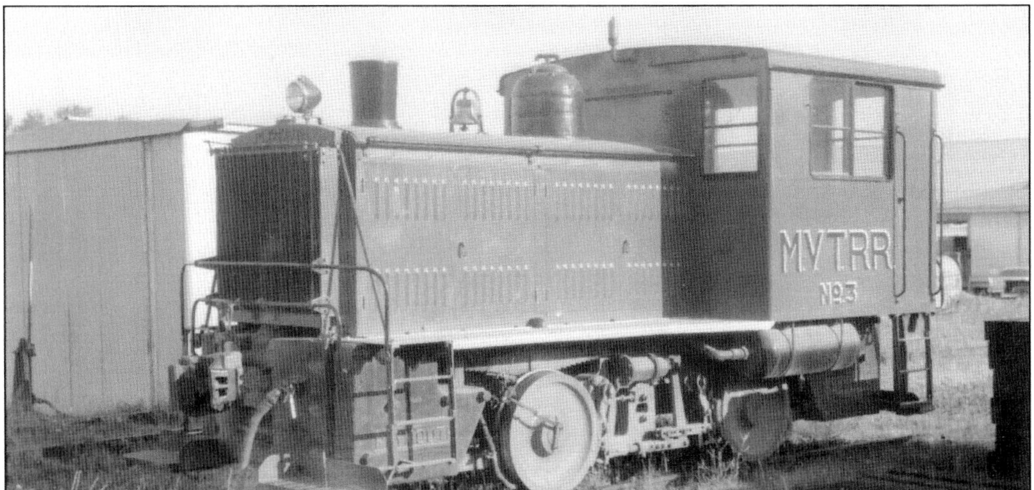

After the failure of the interurban railroad, the tracks in Mount Vernon were acquired in 1933 from Pacific Northwest Traction Company. In 1939, the tracks were abandoned, and the Mount Vernon Terminal Railroad (MVT) was created. For many years, the MVT hauled freight to and from the BNSF Railway's Seattle-Vancouver main line. Operations ceased around 2003, and in March 2005, the company filed to abandon the line. (Courtesy of John Kamb Jr..)

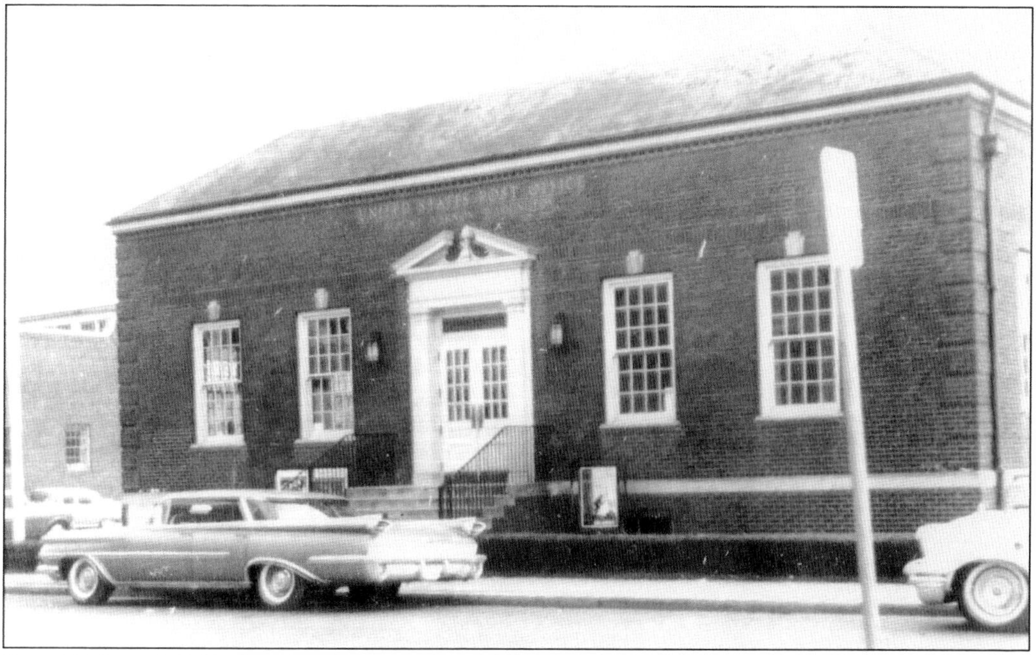

From 1935 until 1964, the Mount Vernon Post Office stood on the corner of Third and Montgomery Streets. The building, pictured around 1960, was constructed as a Works Public Administration (WPA) project. The project also called for a commissioned artwork, and in 1938, artist Ambrose McCarthy Patterson installed the mural *Local Pursuits* (seen below), which features the industries of Mount Vernon and still adorns the upper part of the west lobby wall. Born in 1877 in Daylesford, Victoria, Australia, Patterson came to the Northwest via Paris, Hawaii, and San Francisco. Considered one of the first modern artists to arrive in Seattle, he worked from 1919 until 1947 as an art professor at the University of Washington. The building is currently owned by Skagit Valley College. (Above, courtesy of the Roger Fox collection; below, courtesy of Jessica Bylund.)

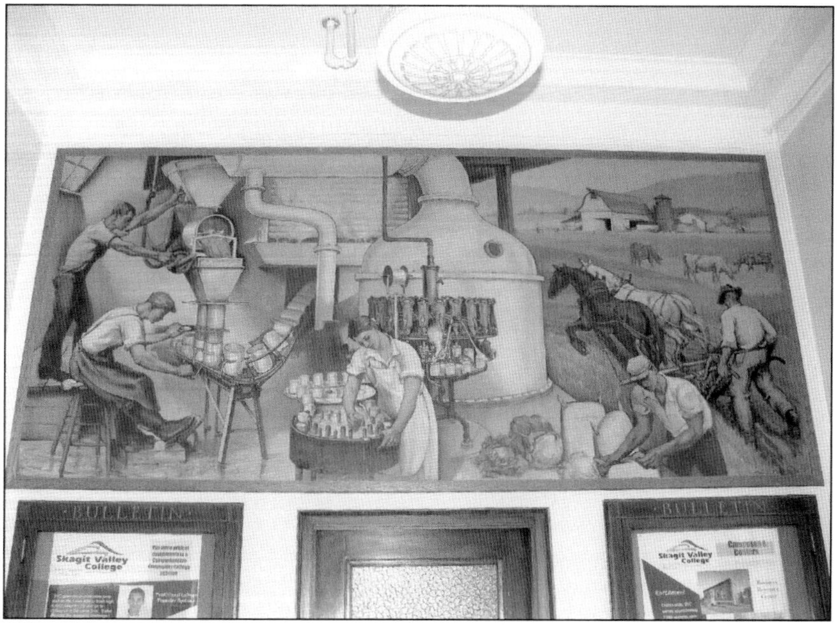

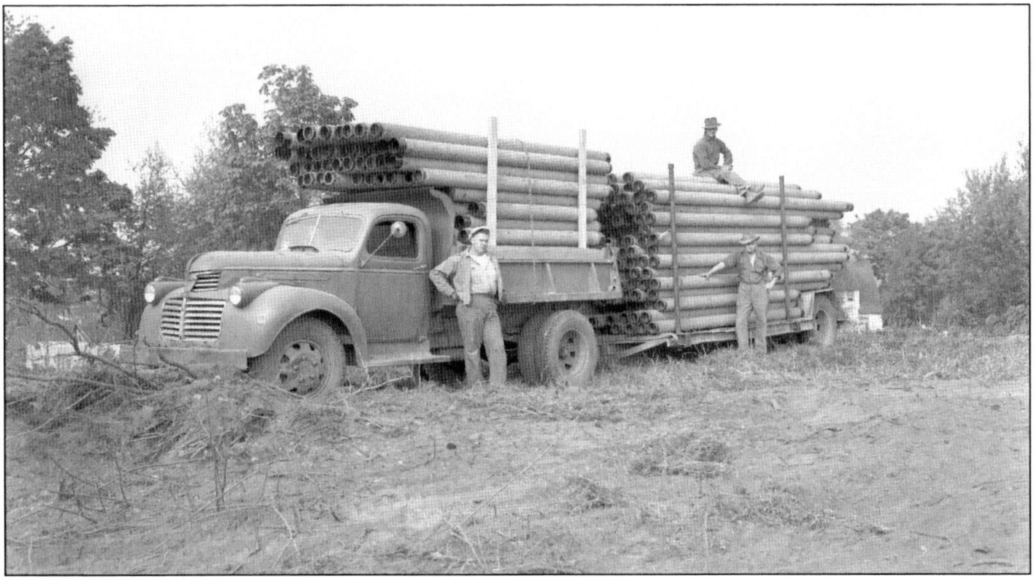

The Skagit County Public Utility District (PUD) was created by the voters of Skagit County in 1936. Within 10 years, the PUD purchased the private water systems that had served the cities of Mount Vernon, Burlington, and Sedro-Woolley, consolidating them into one large system to serve most of the Skagit Valley. Shown here are unidentified PUD workers in 1946. (Photograph by Bill Forman; courtesy of Mark Iverson.)

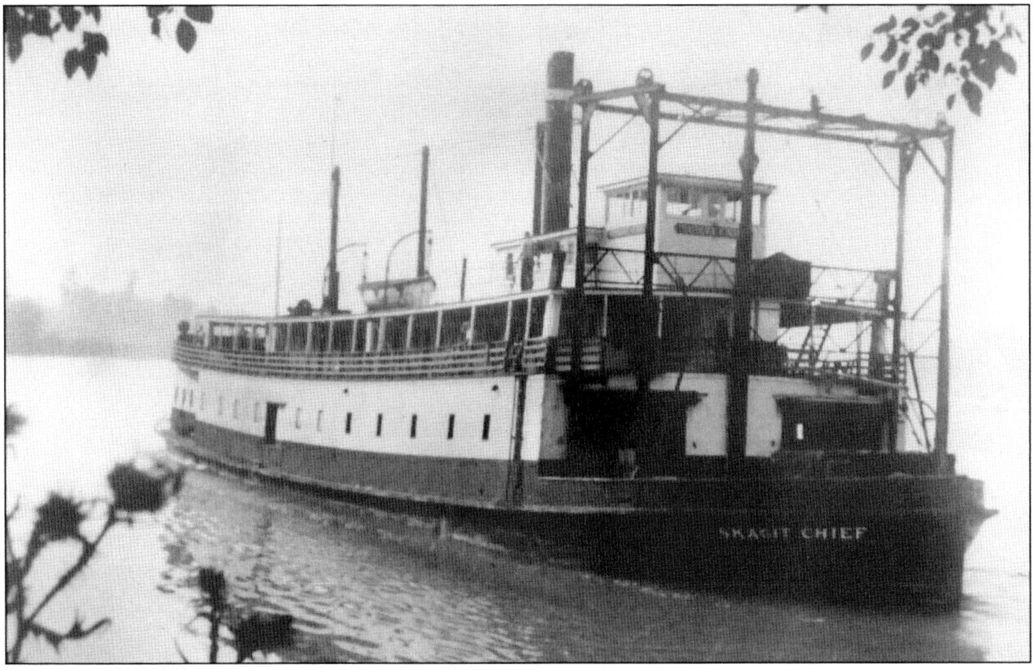

The *Skagit Chief* was a freight stern-wheeler built for the Skagit River Navigation & Trading Company. Built in 1933 by Lake Union Dry Dock & Machine Works for the Seattle, Stanwood, Mount Vernon, and La Conner route, the *Skagit Chief* sank in 1956 while being towed to the Portland Harbor Marina, where it was to be converted into a floating restaurant on the Willamette River. (Courtesy of WSU Skagit County Extension.)

Clint Vaux, older brother of future mayor Harve Vaux, served as mayor from 1935 until 1940. Originally from Minnesota, Vaux came to Mount Vernon and opened a Ford dealership called Vaux Motors. Although a small man, Vaux excelled in both football and track in his youth, earning him the nickname "Dynamite." He later changed careers, selling life insurance and pension plans. (Courtesy of the Mount Vernon Mayor's Office.)

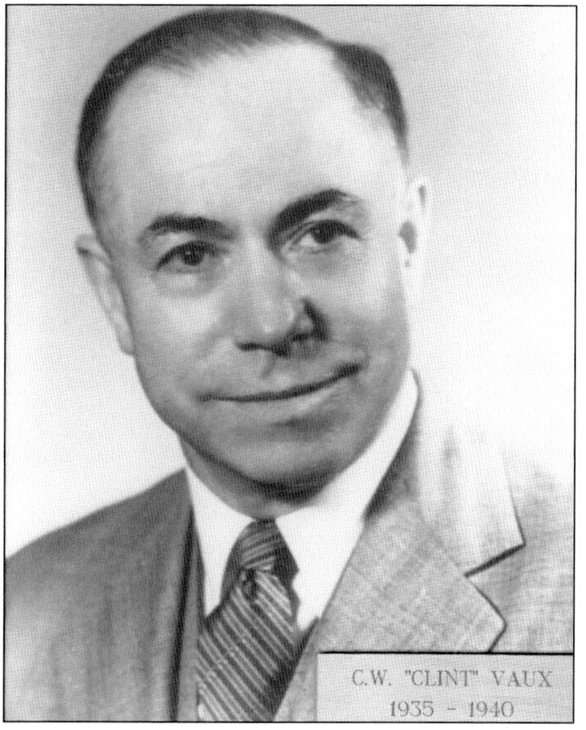

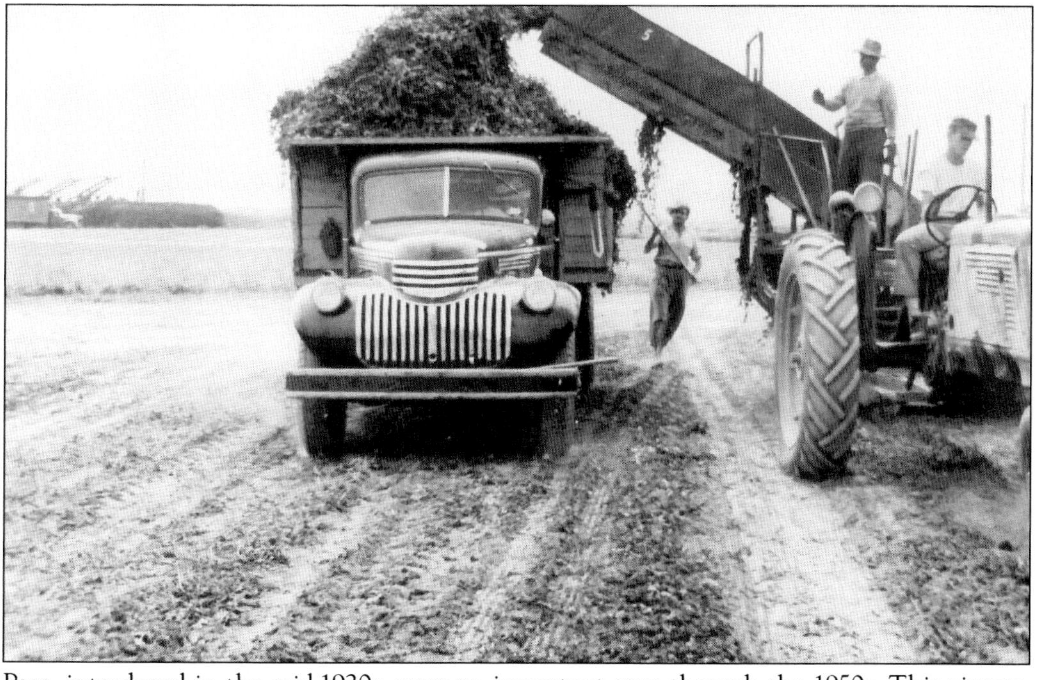

Peas, introduced in the mid-1920s, were an important crop through the 1950s. This picture, taken after World War II, shows the pea-vining process (during the war, women had handled the vining). Peas were processed in town at the frozen food and canning facilities. After the harvest, the vines would ferment and give off a distinctive smell that permeated the air. (Courtesy of WSU Skagit County Extension.)

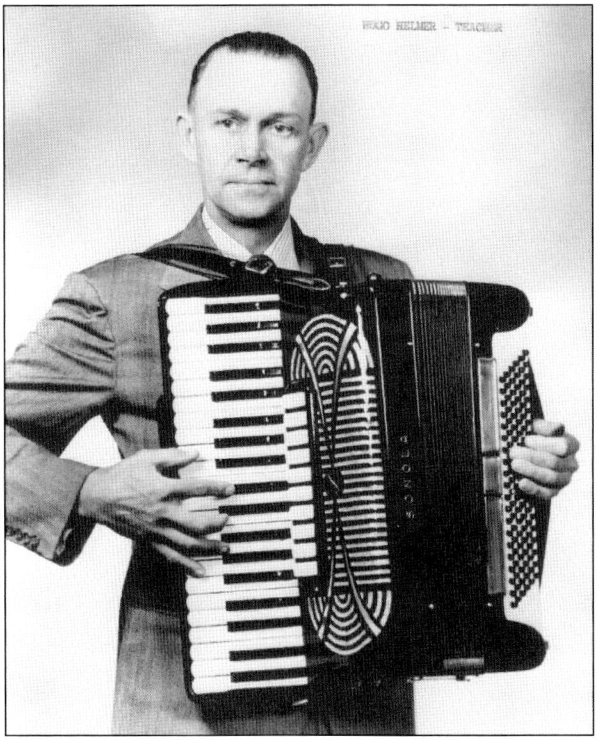

Hugo Gerhard Helmer was born December 23, 1899, in Uddevella, Bohuslan, Sweden, and arrived in Mount Vernon in 1930. In 1935, he started the world's first marching accordion band and in 1937 opened his first music store on Kincaid Street. Hugo Helmer Music later relocated to First Street, occupying the Johnson Building for many years. More than 35 teens and preteens participated in the band, which traveled throughout Washington and parts of British Columbia playing traditional Scandinavian music, polkas, and marches. Below, the band is shown marching through Mount Vernon for the Hickory Hat Days Parade on August 14, 1948. The band ended in 1960, but in 1994, former students and members held a reunion and took up the torch to continue. (Left, courtesy of Ed Marlow; below, photograph by Bill Forman, courtesy of Mark Iverson.)

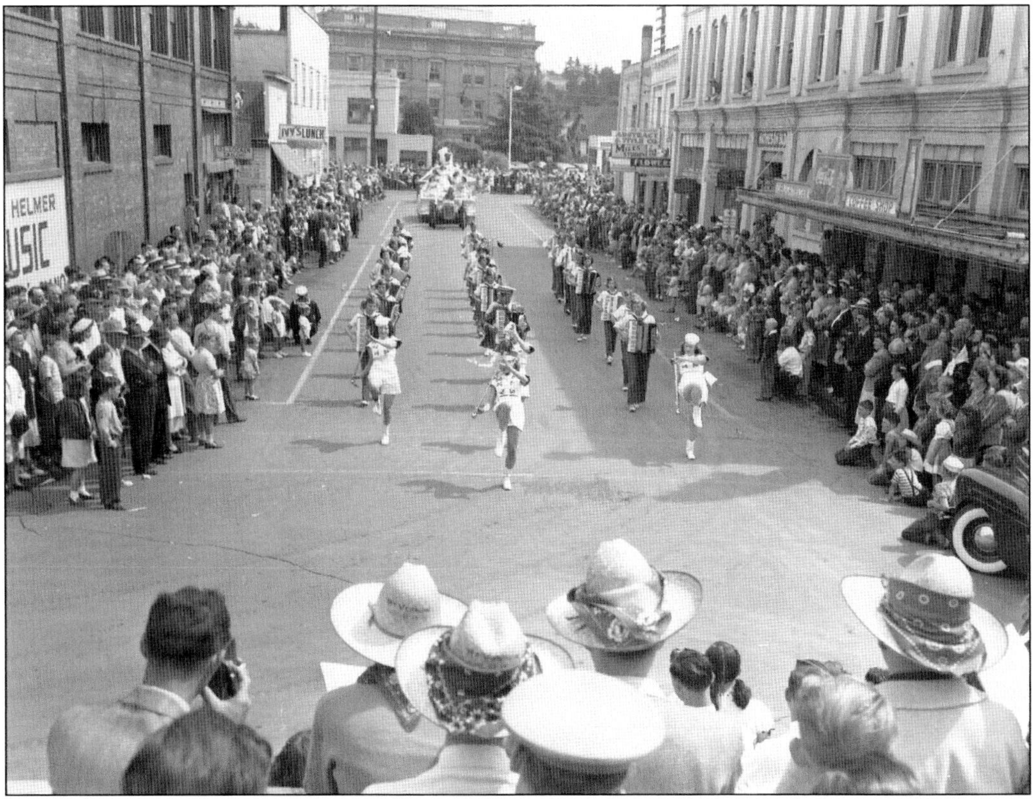

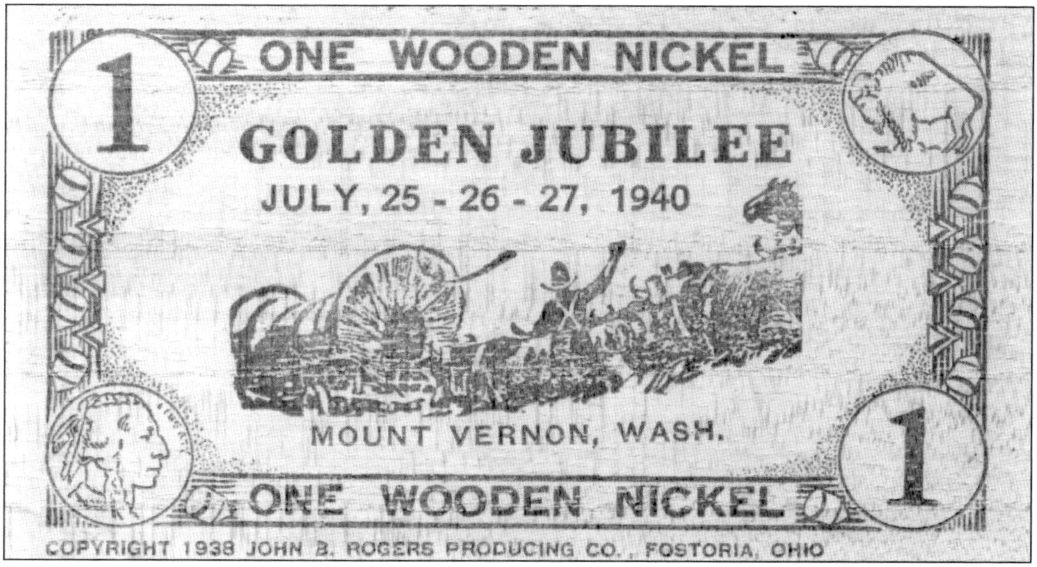

Wooden nickel certificates were issued as souvenirs of the Golden Jubilee Celebration in commemoration of Mount Vernon's 50th anniversary. They were exchangeable in any store in Mount Vernon or were redeemable for cash at any Mount Vernon Bank on or before July 25, 1940. (Courtesy of the Skagit County Auditor's Office.)

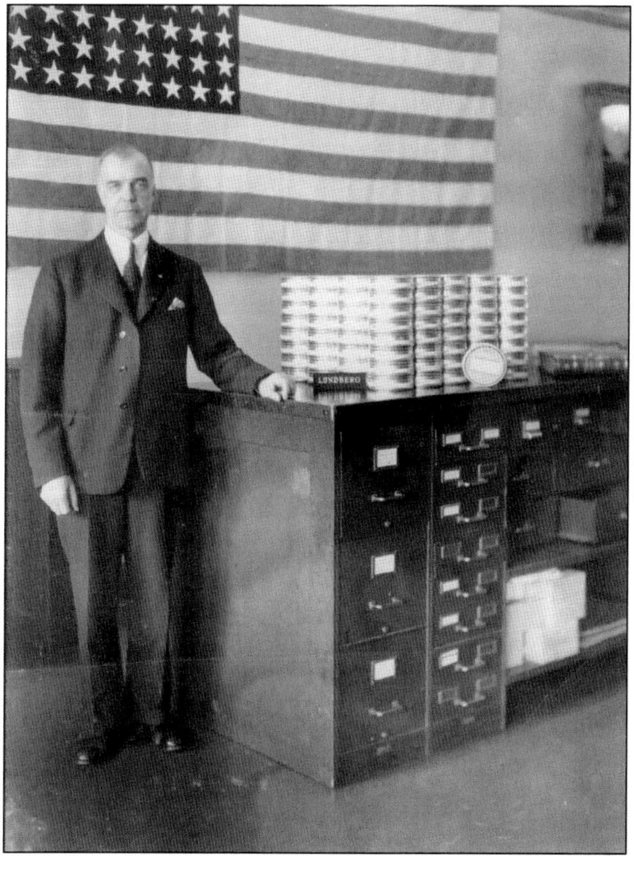

This picture of Carl Lundberg was taken on October 1, 1942. Lundberg was hired to archive Skagit County's records on microfilm, and the back of the photograph states, "These containers hold all of the essential records of the courthouse in Skagit County, Washington. 488,310 pages, including maps and files, are duplicated on six and one third miles of non-inflammable micro-film." (Courtesy of the Skagit County Auditor's Office.)

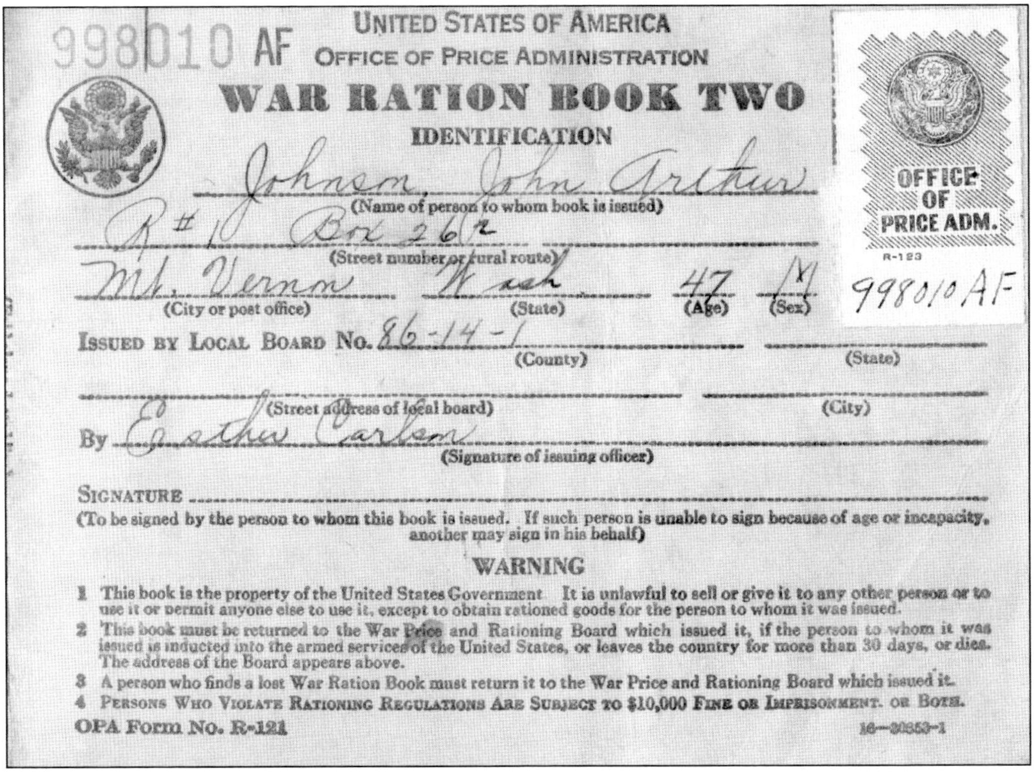

998010 AF

UNITED STATES OF AMERICA
OFFICE OF PRICE ADMINISTRATION

WAR RATION BOOK TWO

IDENTIFICATION

Johnson, John Arthur
(Name of person to whom book is issued)

R #1 Box 26 1/2
(Street number or rural route)

Mt. Vernon Wash 47 M
(City or post office) (State) (Age) (Sex)

OFFICE OF PRICE ADM.
R-123
998010 AF

ISSUED BY LOCAL BOARD NO. 86-14-1
(County) (State)

(Street address of local board) (City)

By Esther Carlson
(Signature of issuing officer)

SIGNATURE
(To be signed by the person to whom this book is issued. If such person is unable to sign because of age or incapacity, another may sign in his behalf)

WARNING

1 This book is the property of the United States Government It is unlawful to sell or give it to any other person or to use it or permit anyone else to use it, except to obtain rationed goods for the person to whom it was issued.
2 This book must be returned to the War Price and Rationing Board which issued it, if the person to whom it was issued is inducted into the armed services of the United States, or leaves the country for more than 30 days, or dies. The address of the Board appears above.
3 A person who finds a lost War Ration Book must return it to the War Price and Rationing Board which issued it.
4 PERSONS WHO VIOLATE RATIONING REGULATIONS ARE SUBJECT TO $10,000 FINE OR IMPRISONMENT, OR BOTH.

OPA Form No. R-121 16—30853-1

Ration books, like this one from January 1943, were issued by the Office of Price Administration (OPA) beginning in the spring of 1942. They contained stamps and gave precise details about the amounts and types of food recipients were allowed. Four different series were issued; by the end of the war, over 100 million copies of each ration book had been printed. (Courtesy of the Basye family.)

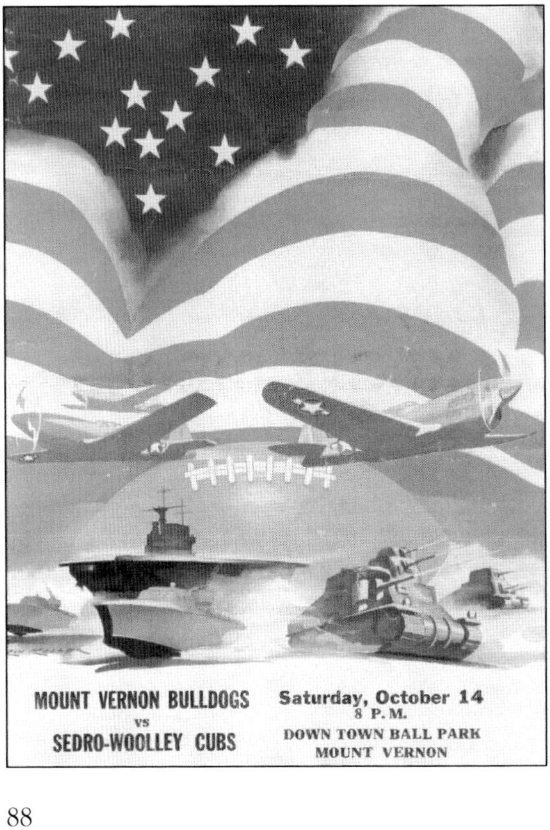

Patriotism abounded in Skagit County. This 1944 flier advertises a football game between the Mount Vernon Bulldogs and the Sedro-Woolley Cubs. Although the game ended with a Mount Vernon defeat, the 1945 *Skagina* declared the 18-12 loss "easily the Bulldog's best game of the season." It was the team's first season under coach Royal E. "Pop" Gunn. (Courtesy of the Mount Vernon Mayor's Office.)

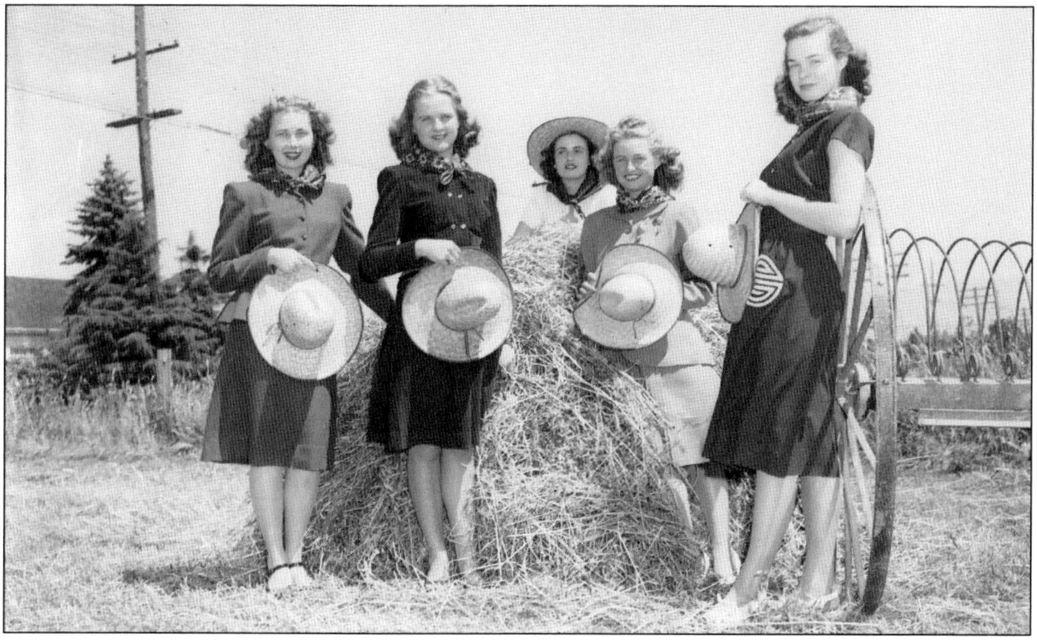

In the 1930s, the annual Mount Vernon Pow-Wow transformed into Hickory Hat Days, which continued into the 1950s. The three-day community pride celebration was held during the fourth week of July until the 1940s, when it began occurring in July or August. Thousands of visitors came to Mount Vernon for the event. Festivities, which featured special hats and neckerchiefs, included parades, street dances, kangaroo courts, baseball games, a carnival, and beauty queens. The picture above shows the contestants for Hickory Hat Queen in 1947. From left to right are Dorothy Hanson, Leona Crandall, Patricia Sayre, Betty Lockhart (the eventual queen), and Anne Marie Looy. The picture below shows unidentified festival officials in August 1948. (Both, photograph by Bill Forman; courtesy of Mark Iverson.)

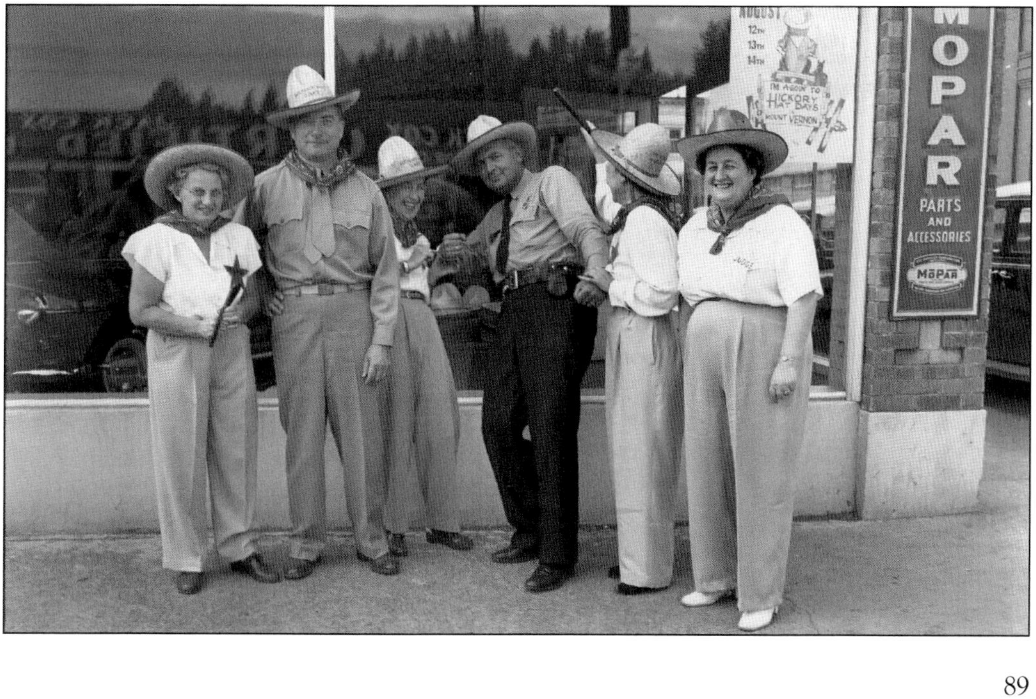

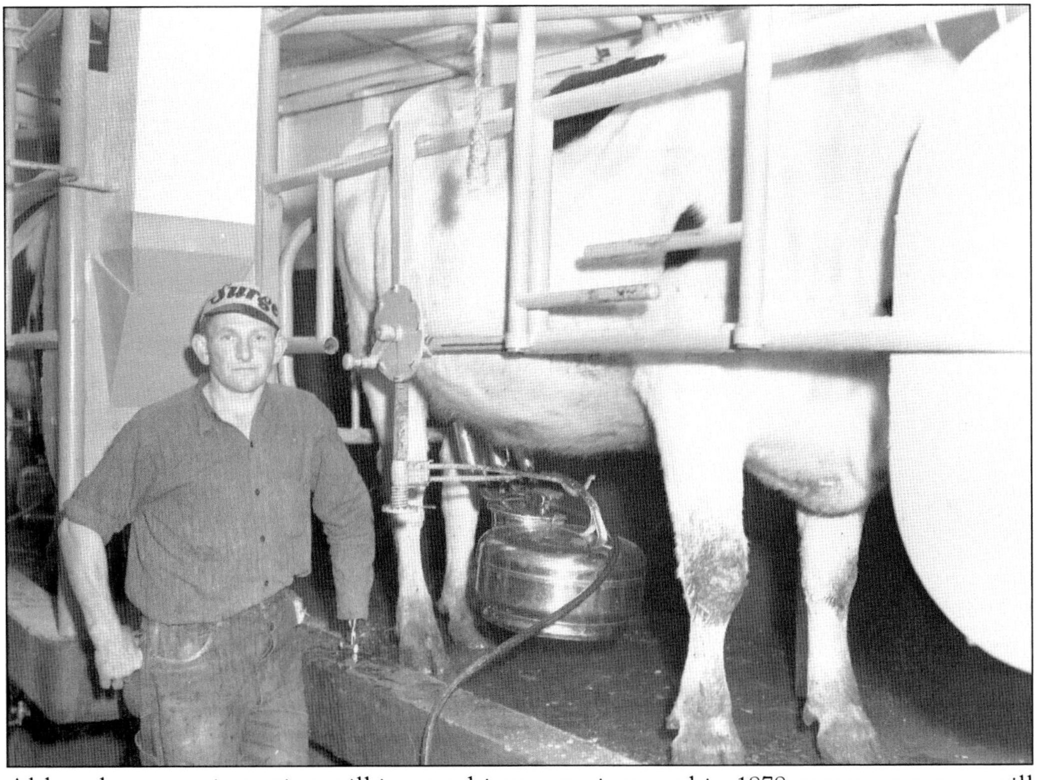

Although automatic suction milking machines were invented in 1878, many cows were still milked by hand as late as the 1940s. This local dairy farmer and automatic milking equipment were photographed in 1948. The necessity of new, more advanced technology contributed to the rising costs that dairy farmers faced. (Photograph by Bill Forman; courtesy of Mark Iverson.)

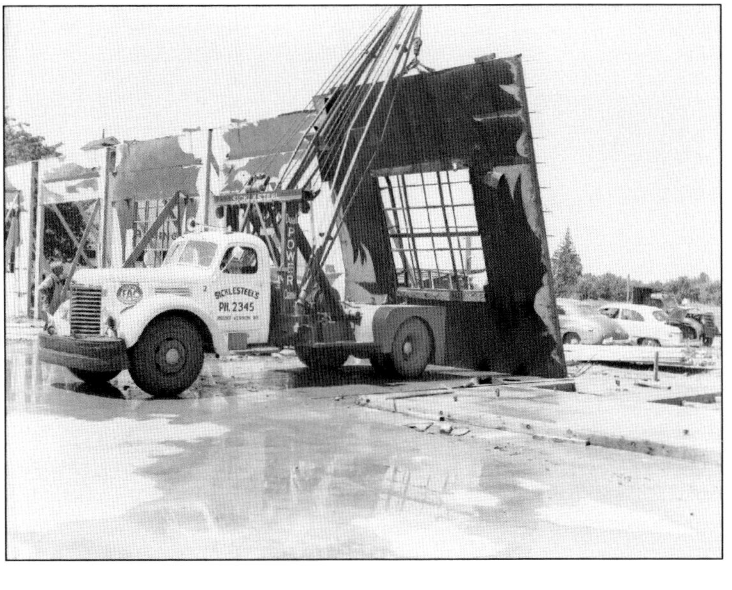

Still in business today as Sicklesteel Cranes Inc., Sicklesteel Motor Company was founded in 1937. Their first tow truck was a converted 1929 Cadillac V-12 roadster and offered 24-hour towing. In 1949, Sicklesteel set the first precast-concrete tilt-up building in Skagit County with a tow truck, similar to the one seen here in 1950. In 1952, Sicklesteel purchased their first crane. (Photograph by Bill Forman; courtesy of Mark Iverson.)

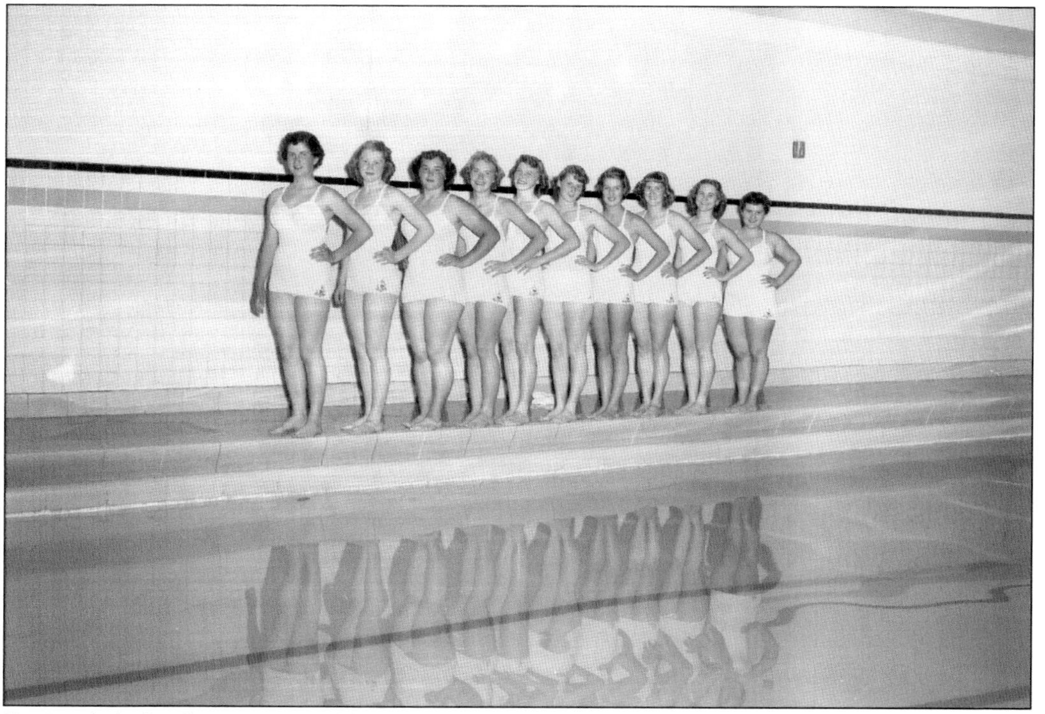

Ilo Sande helped organize and build the YMCA swimming pool as part of their new facility, opened in 1941. Mount Vernon's YMCA facility was funded partially by a generous 1920s monetary donation and was constructed on land belonging (even today) to the school district. In exchange for payment, the YMCA agreed to let the high school use the pool for their aquatic sports. In 1946, Mount Vernon's first swim team started, under the direction of Coach Sande. Sande was a 2005 inductee into the Washington Interscholastic Swim Coaches Association Hall of Fame for being a pioneer in state high school swimming history. Another group that relied on the new pool was the high school's Flamingo Club, pictured (poolside and in action) in 1951. (Both, photograph by Bill Forman; courtesy of Mark Iverson.)

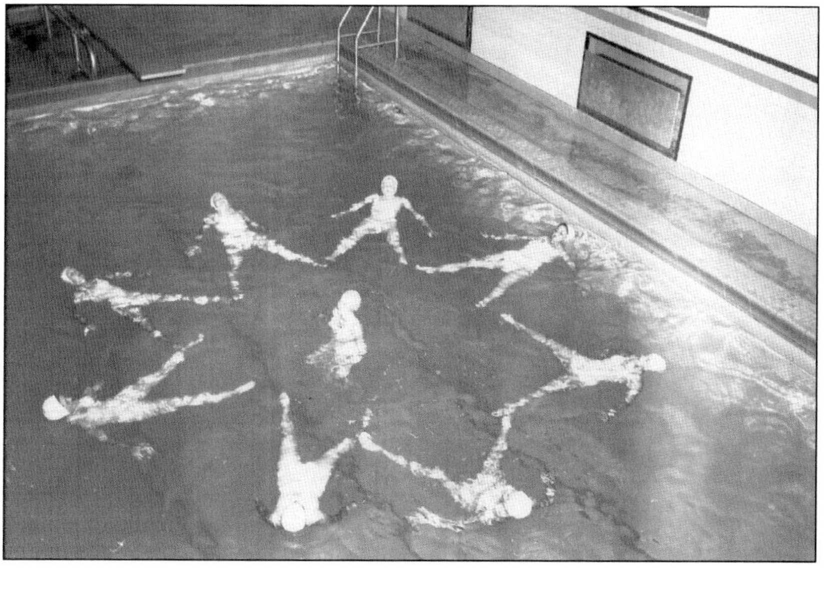

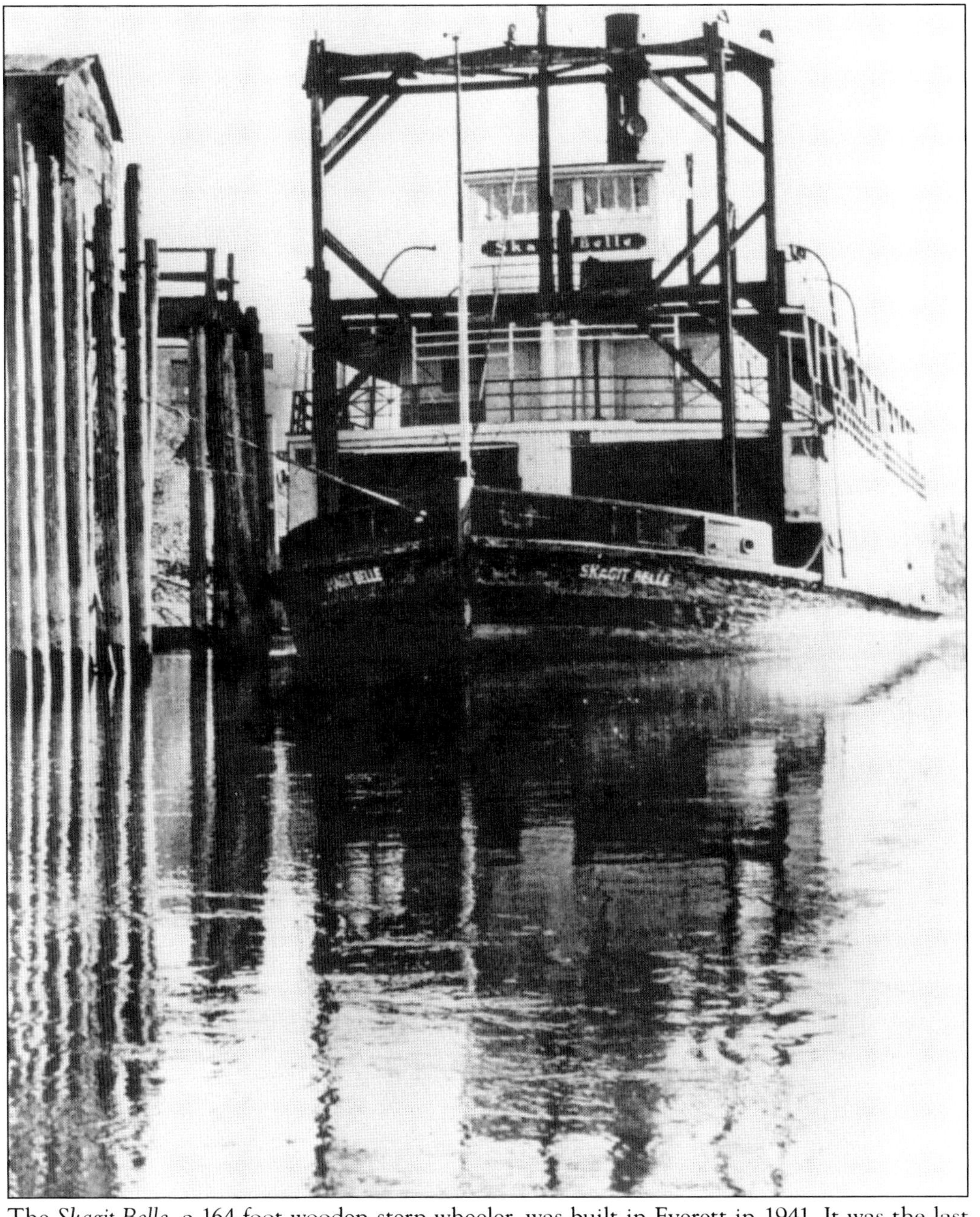

The *Skagit Belle*, a 164-foot wooden stern-wheeler, was built in Everett in 1941. It was the last commercial stern-wheeler steamboat on Puget Sound. Shortly after manufacture, the vessel was requisitioned for war service and was not returned to the Skagit River Navigation Company until 1947. In 1950, the *Skagit Belle* joined the *Skagit Chief* and the *W.T. Preston* in a race of stern-wheelers for Seattle's inaugural Seafair. The *W.T. Preston* won the race handily. The *Skagit Belle* continued as a freighter until sustaining damage on a sandbar in 1959. It was repaired in Bellingham before going to Seattle and being parked next to Pier 51, where it was used as a floating restaurant during the 1962 World's Fair. In 1965, the *Skagit Belle* sprung a leak and sank. It was not until 1973 that the boat was dredged up and hauled away for disposal. (Courtesy of Ed Marlow.)

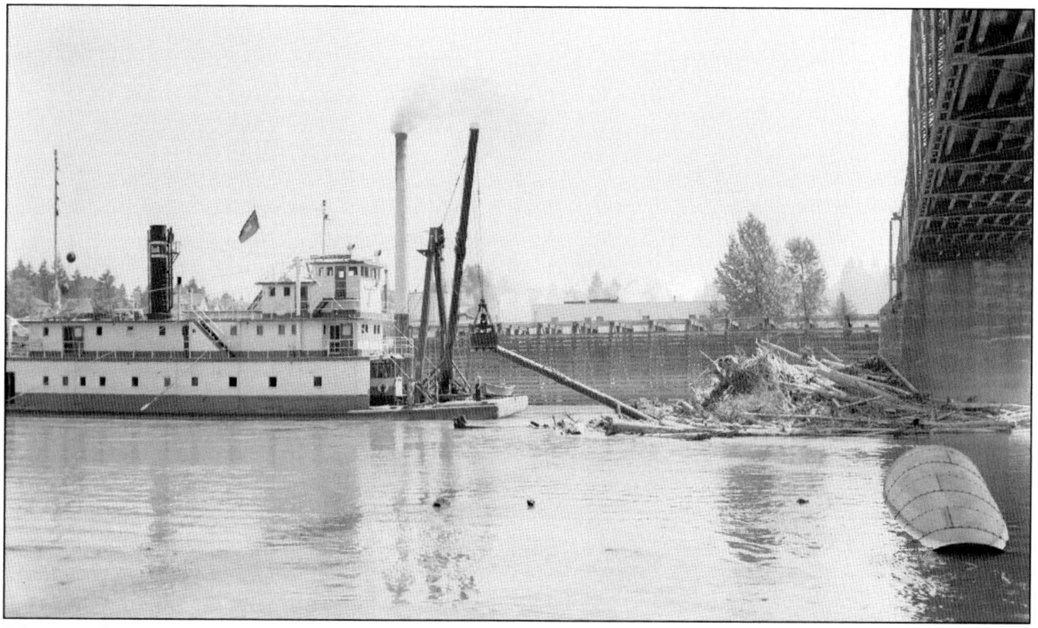

The snagboat *W.T. Preston* clears debris from the Skagit River in August 1946. The *W.T. Preston* was one of the last stern-wheelers to work in Puget Sound and is one of only two snagboats remaining in the contiguous United States. It was designated a National Historic Landmark in 1989 and is now located at the Anacortes Museum & Maritime Heritage Center. (Photograph by Bill Forman; courtesy of Mark Iverson.)

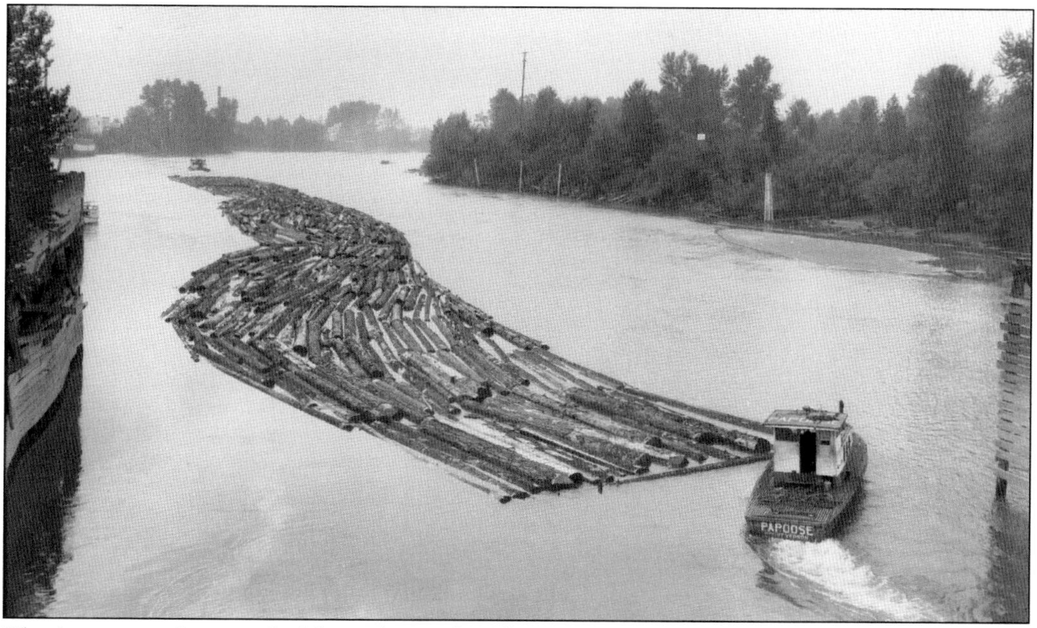

The logging industry relied heavily on the Skagit River to transport timber for many years. Tugboats like the *Papoose*, working in 1946, were entrusted with large log booms as they traveled downstream toward a mill. In time, it became more efficient to transport logs on trains and trucks. Additionally, environmental concerns curtailed the few remaining river transportation routes for timber. (Photograph by Bill Forman; courtesy of Mark Iverson.)

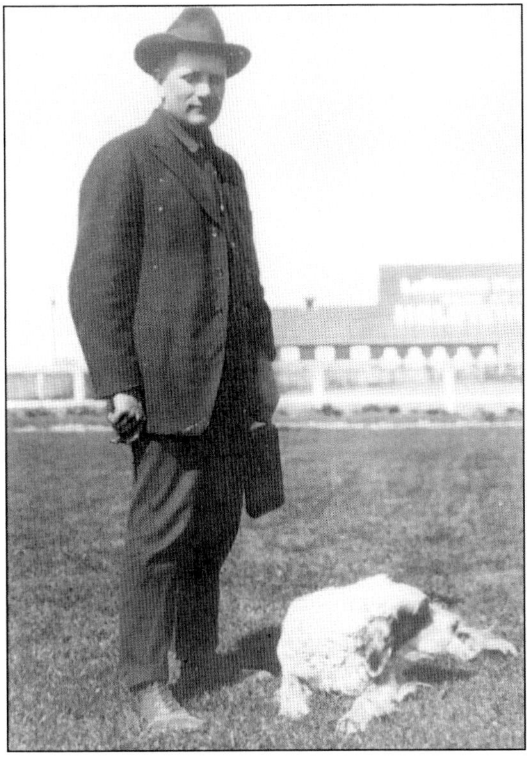

James Hulbert began growing vegetable seed in the 1920s. When a virus threatened to destroy the local seed crops, he was instrumental in bringing agricultural scientists to Mount Vernon in the early 1940s. As a result, crop isolation and rotation procedures were implemented, which saved the area's seed industry. Hulbert was also known for never being without a cigar. (Courtesy of Colleen Hulbert Fischer.)

Dr. George C. Folger served as president of the Washington State Veterinary Medical Association in 1946. Having graduated from the State College of Washington (now WSU) in 1937 with a degree in veterinary medicine, he came to work in Mount Vernon with Dr. C.S. Phillips, who had been one of only two graduates from WSU's first veterinary medicine class in 1902. (Photograph by Bill Forman; courtesy of Mark Iverson.)

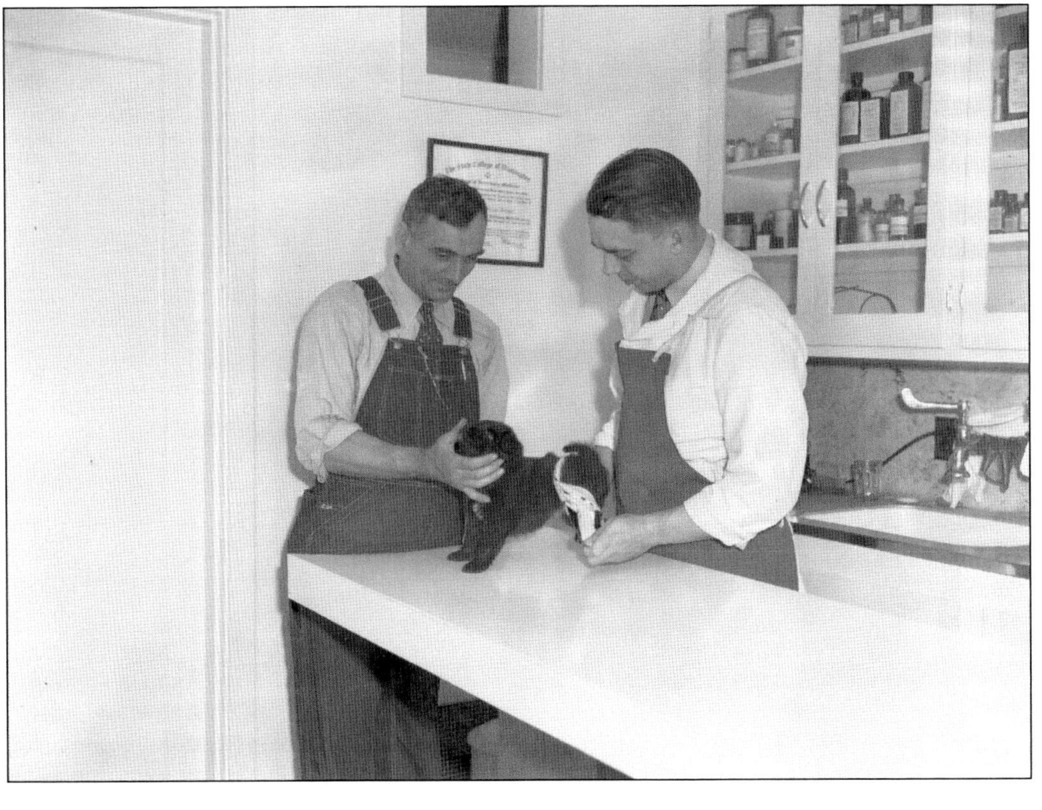

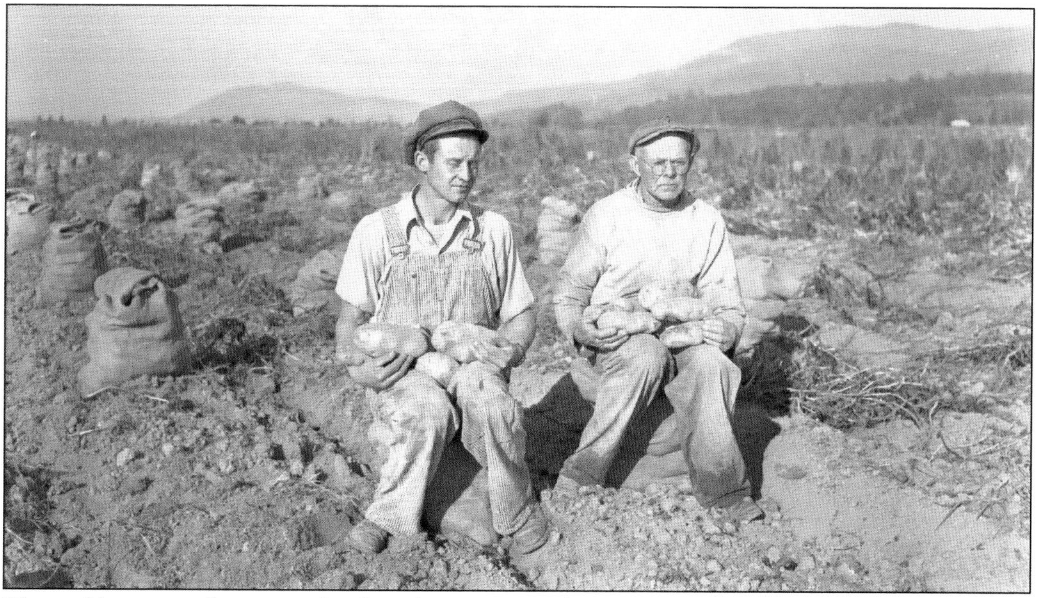

The mild marine climate and rich soil create a perfect environment for red, white, and yellow potatoes. Seen here in 1946 are two unidentified farmers with fine examples of the potatoes grown in Mount Vernon. Valuable cash crops, like potatoes, helped pull Mount Vernon out of the Depression and remain an important part of the local economy to this day. (Photograph by Bill Forman; courtesy of Mark Iverson.)

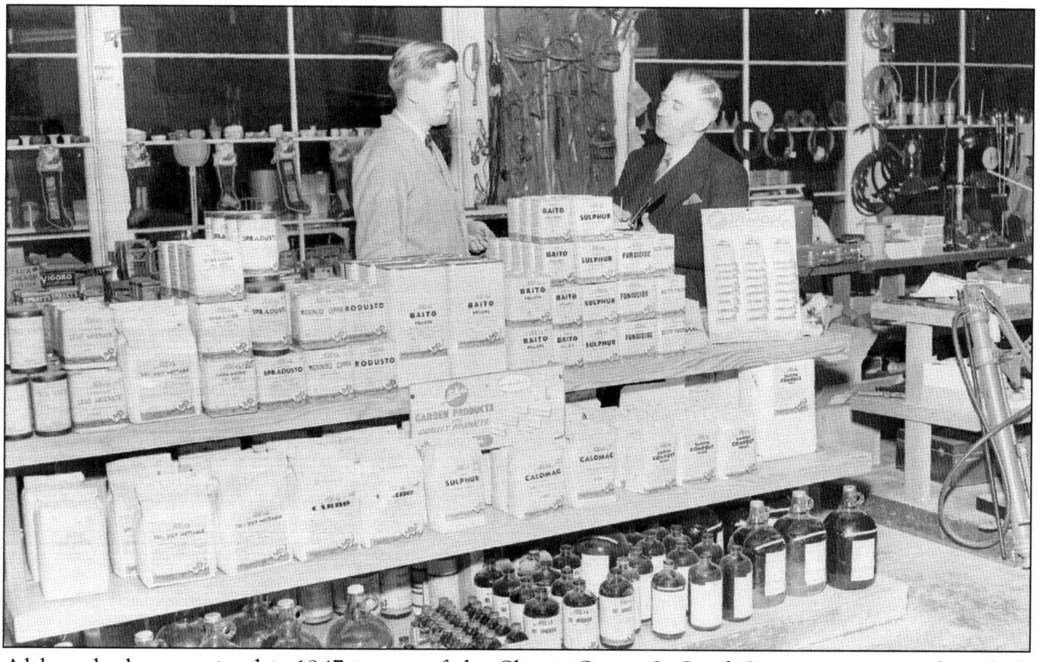

Although the men in this 1947 image of the Skagit Grain & Seed Company are unidentified, it is interesting to note the abundance and variety of pesticides on the shelves. It was claimed that the DDT Dairy Barn Spray, seen in the bottom left, would "improve the flow of milk by 15 percent" by killing flies. The use of DDT was banned in 1973. (Photograph by Bill Forman; courtesy of Mark Iverson.)

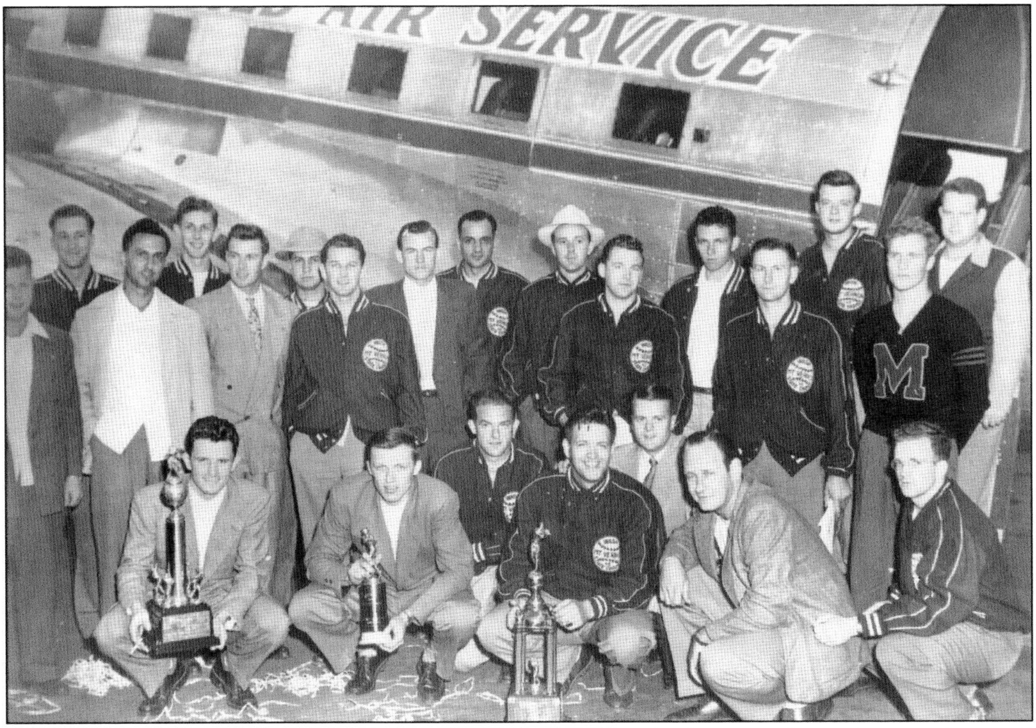

On August 2, 1948, Mount Vernon's Milkmaids won the Washington State Semiprofessional Baseball Championship by defeating the defending champions, the Bellingham Bells, 2-1. The victory earned the Milkmaids (pictured above) a berth in the National Semiprofessional Tournament in Wichita, Kansas, where they placed third. Catcher Sammy White, a University of Washington alum who went on to play with the Red Sox, was named the outstanding player of the tournament. In 1949, the team returned to the National Tournament as one of the top-ranked teams but was upset early by the Weimer Texas Truckers. Below is a photograph of the 1950 team. (Above, courtesy of the Roger Fox collection; below, photograph by Bill Forman, courtesy of Mark Iverson.)

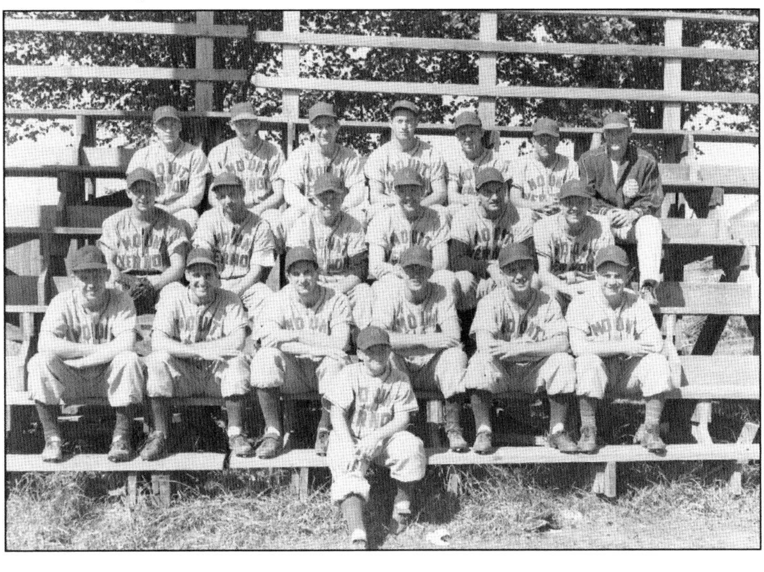

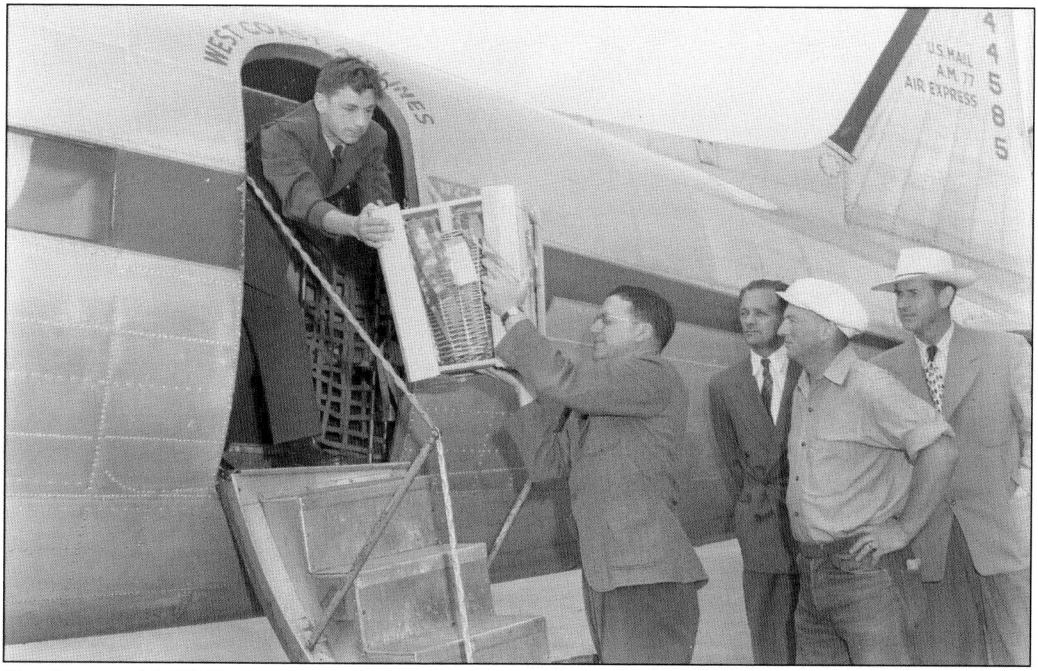

On August 16, 1949, the City of Mount Vernon sent a large gold-finished hamper filled with Skagit Valley tulip, iris, and daffodil bulbs to the citizens of Wichita, Kansas, for planting in their parks. Mayor Harve Vaux placed the gift aboard the plane. The gesture was made in advance of the Milkmaids going to Wichita to compete for the national semiprofessional title. (Photograph by Bill Forman; courtesy of Mark Iverson.)

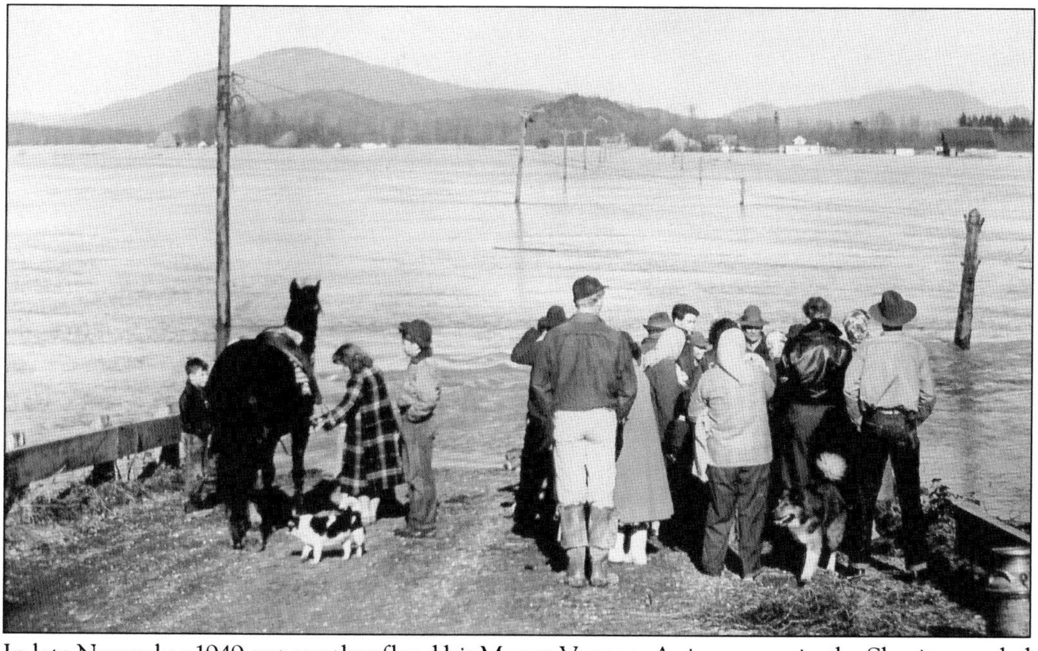

In late November 1949, yet another flood hit Mount Vernon. A river gauge in the Skagit recorded a historic crest of 34.21 feet on November 18. Here, people and animals survey the magnitude of flooding on Francis Road. (Courtesy of WSU Skagit County Extension.)

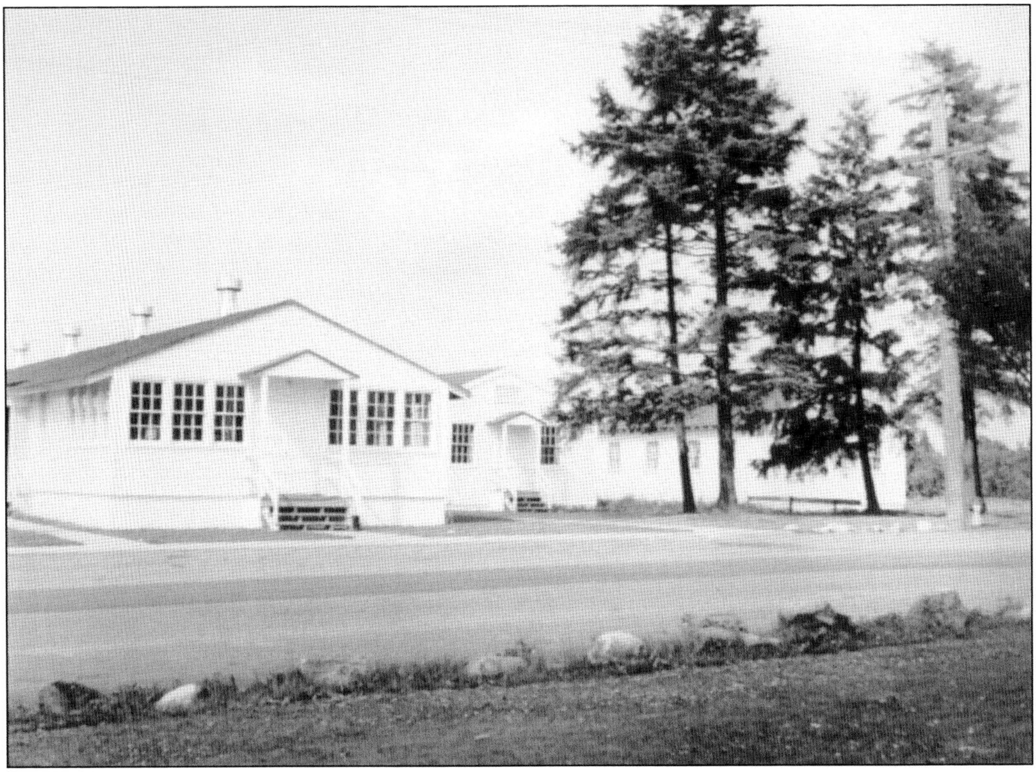

In 1945, the junior college was awarded $330,000 from the Washington State Legislature, which it used to purchase a five-acre tract of land adjacent to the high school. Surplus government buildings from Paine Field were moved onto the new land to accommodate the students, while administrative offices and laboratory space remained in the high school. The college remained here until the late 1950s. (Courtesy of the Mount Vernon Mayor's Office.)

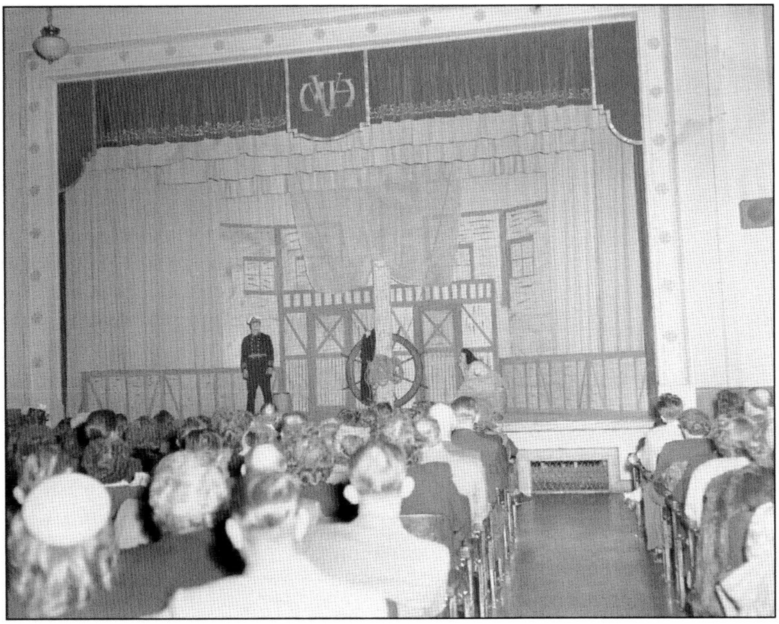

This is a scene from an April 22, 1948, performance of Gilbert and Sullivan's *H.M.S. Pinafore*. The production was sponsored by the junior chamber of commerce and starred an all-local cast. Margaret Isling (right) played Josephine, the captain's daughter. The high school auditorium hosted the show. (Photograph by Bill Forman; courtesy of Mark Iverson.)

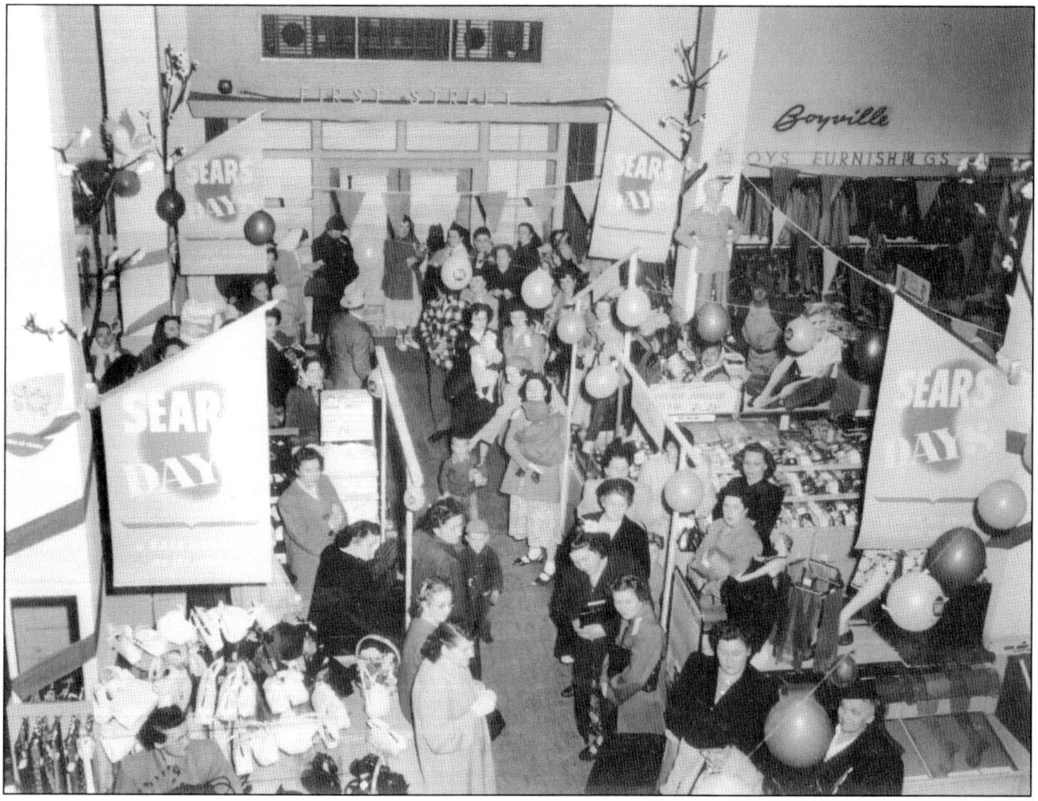

Downtown was still the main commercial district in the 1940s. In addition to local stores, offices, and theaters, there were also national chain department stores. J.C. Penney came to Mount Vernon in 1923 and expanded with a two-story addition in 1949. Sears had a grand opening (pictured above) on March 4, 1947. The large store, located on First Street, was a mainstay for many years. Both Sears and J.C. Penney are visible in the photograph of First Street, below. As the city expanded over the years, the large, downtown department stores became inconvenient and lost business to the newer suburban shopping malls. (Above, courtesy of the Mount Vernon Mayor's Office; below, courtesy of the Roger Fox collection.)

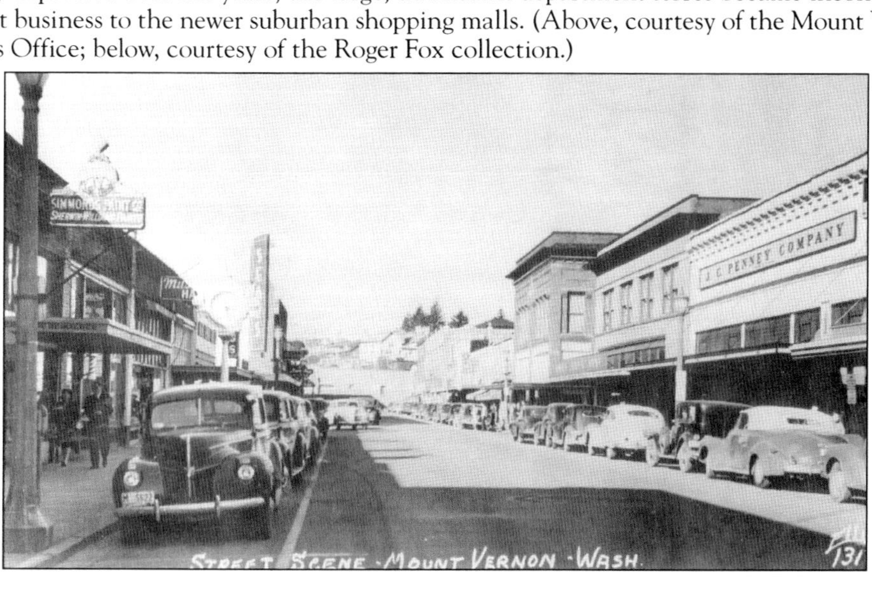

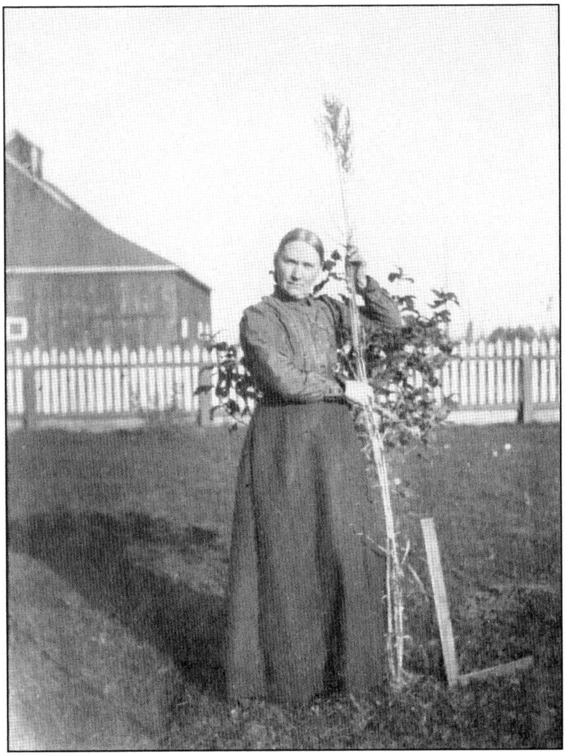

Britta Johnson is seen planting a poplar tree at her farm on Beaver Marsh Road. Lombardy poplars, native to the Mediterranean region, became very fashionable in America. The trees, which form a slender column of many short, upward-pointing branches, can reach between 40 and 60 feet in height, with a spread of only 10 to 12 feet. They were prized for their quick growth rate and ability to serve as shelter or shade in the Skagit Valley flats. Their preference for moist, rich soil also made them good trees for the valley. Unfortunately, they are susceptible to canker disease and do not have a very long lifespan. (Courtesy of the Bayse family.)

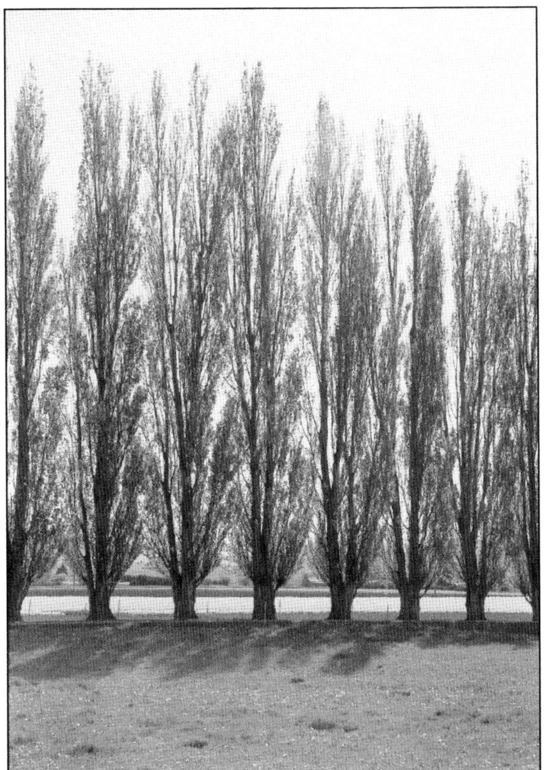

Today, Lombardy poplars, coupled with blooming tulips and daffodils, provide a prized photographic opportunity every spring. (Courtesy of David Hock.)

Seven

MODERNIZATION 1950–1979

The 1950s were a time of widespread expansion for Mount Vernon. New schools, such as Washington (1950), Madison (1954), and Jefferson (1956), were built. Additionally, substantial improvements were made to old schools—a new gymnasium (1950) and agriculture building (1952) were added at the high school, and an addition was made to Washington, only a few years after construction, in 1953. The Riverside area was annexed to Mount Vernon in 1954. The junior college purchased a 35-acre site in northeast Mount Vernon in 1957 and opened a campus there in 1959. But the biggest change of this decade was the creation of Skagit Valley Hospital. Fundraising efforts had begun in 1947, and the hospital district was established in 1954. Land was purchased between 1954 and 1955, and the two-story community hospital opened in 1958. Credit is due to the initiative of citizens and a group called the Women's Guild for making it all possible. Today, the hospital is one of the largest employers in Mount Vernon.

In the 1960s, Interstate 5 replaced Highway 99, sometimes eclipsing the old road, at other times, paralleling it. The new freeway brought more people to the valley and cemented Mount Vernon's links with the larger cities of Everett, Seattle, and Bellingham. Unfortunately, it also enhanced the existing separation between downtown and "the Hill." This led Mount Vernon and Skagit County to realize the need to establish the direction and character of growth through comprehensive planning. John Graham completed Mount Vernon's first Comprehensive Plan in 1960. Skagit County followed in 1963 with a plan of its own, which resulted in zoning in 1966.

Mount Vernon's population began to boom in the 1970s, expanding from 8,800 to more than 13,000. During that time, the city annexed 2.32 square miles of county land to the north and east. This launched a conversion of agricultural lands to new commercial and residential zones. The construction of the Mount Vernon Mall (1971) and the Skagit Valley Mall (1973) created a major shift in the dynamic of the city, and many downtown businesses began to suffer.

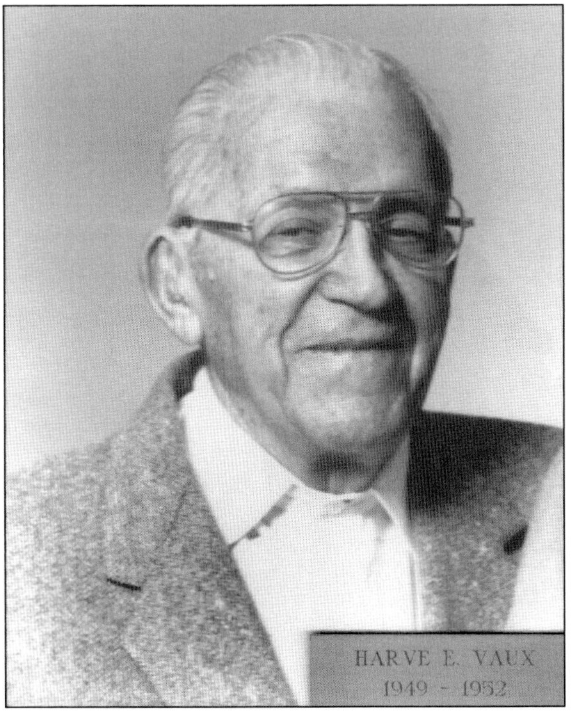

Harve E. Vaux was the mayor of Mount Vernon from 1949 until 1952. For many years, Harve worked at his older brother Clint's Ford dealership. After the dealership was sold, Harve started his own successful tractor-sales business. By importing English Ferguson tractors, he was able to undercut the rest of the tractor market and do brisk business throughout the farming community. (Courtesy of the Mount Vernon Mayor's Office.)

Seen here in the early 1950s, Skagit County's wagon drill blasts rock for revetment purposes in the city of Mount Vernon's new quartzite quarry. The men, who were city or county employees, are, from left to right, ? Henry, Dave Sawyer, L. Brendle (blasting foreman), and Ted Schnider. (Courtesy of the Roger Fox collection.)

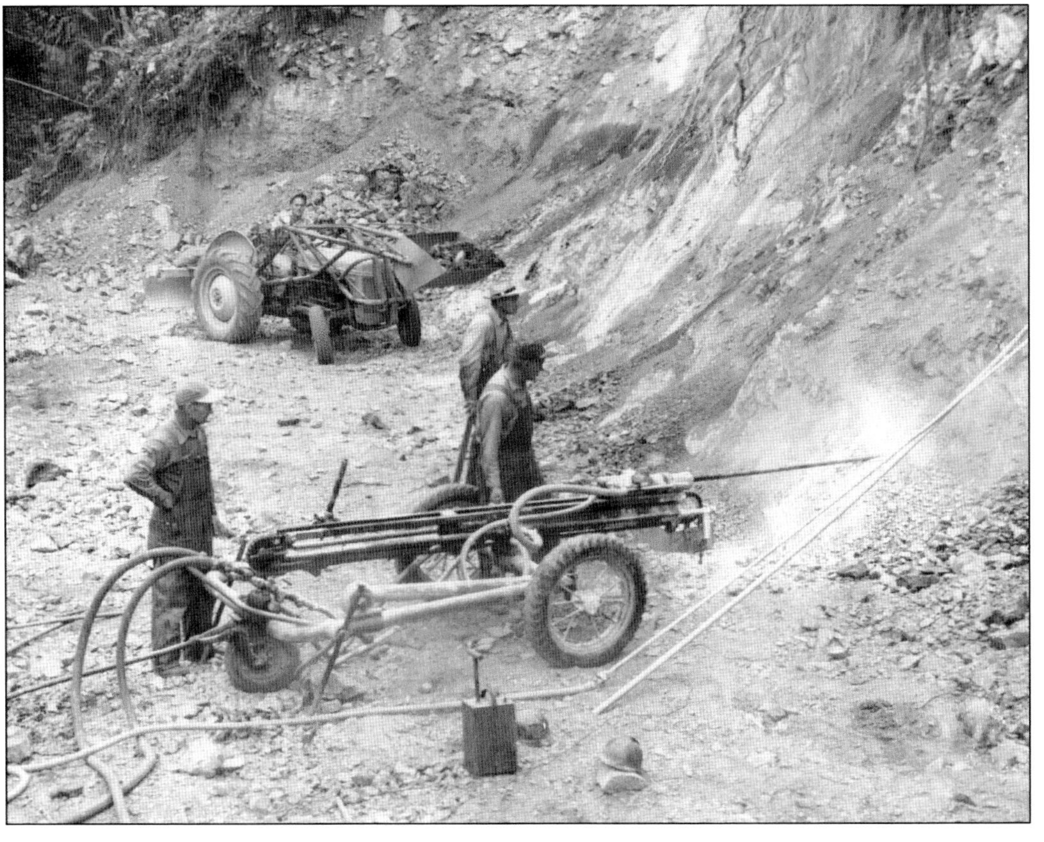

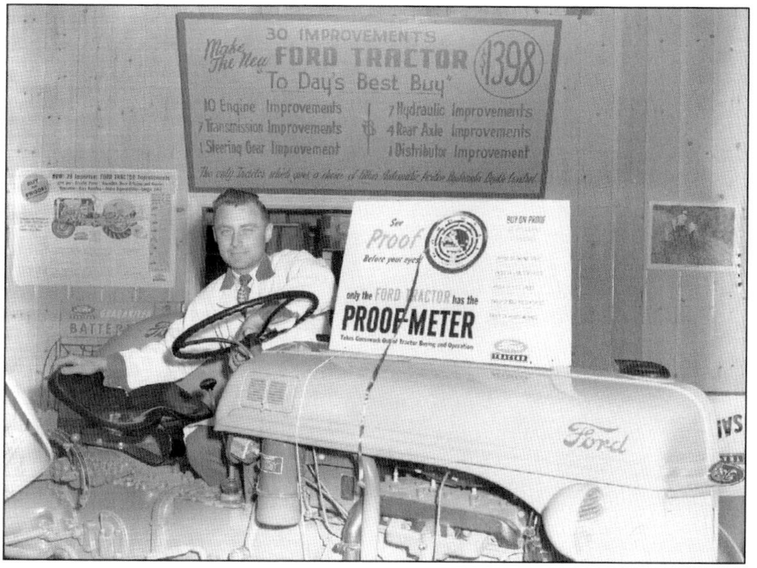

This 1950 image shows a local Ford dealer with a Model 8N tractor (note the price: $1,398). In the 1950s, Ford was battling a lawsuit with former partner Harry Ferguson as well as trying to rebuild their market share after being out of the American tractor business during the 1930s and surviving wartime restrictions in the 1940s. (Photograph by Bill Forman; courtesy of Mark Iverson.)

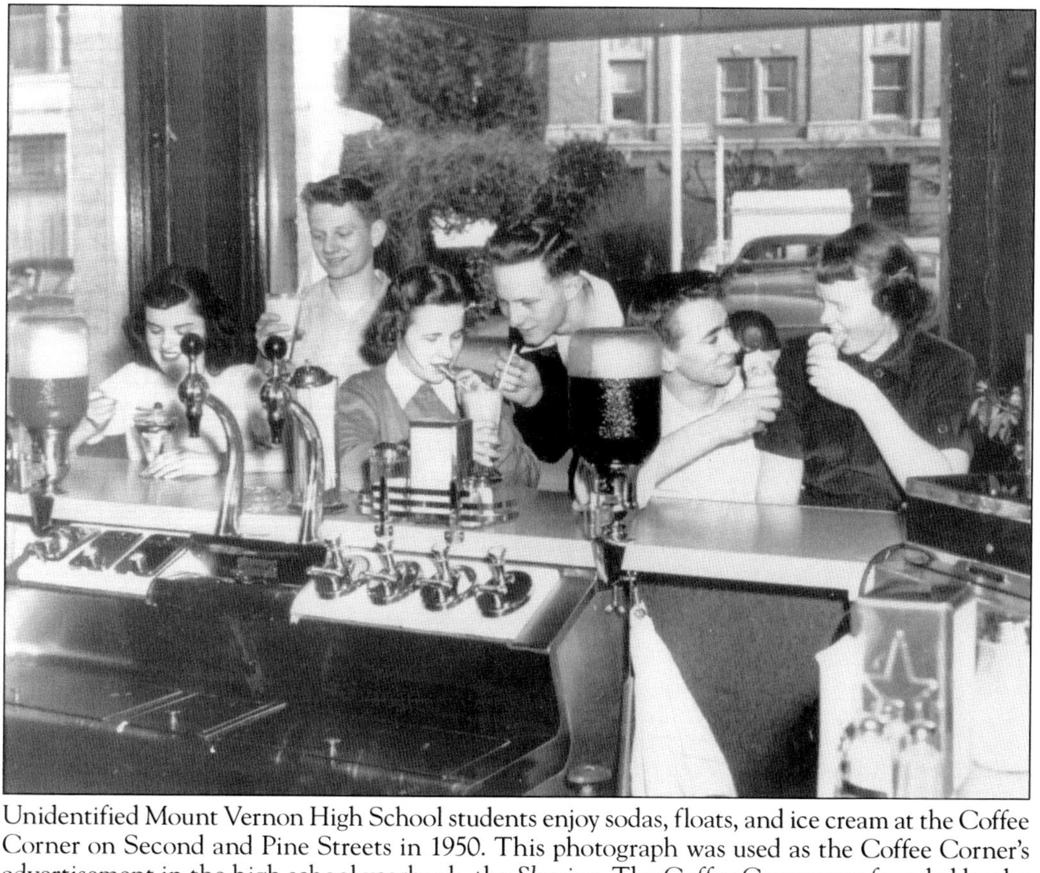

Unidentified Mount Vernon High School students enjoy sodas, floats, and ice cream at the Coffee Corner on Second and Pine Streets in 1950. This photograph was used as the Coffee Corner's advertisement in the high school yearbook, the *Skagina*. The Coffee Corner was founded by the Hansons in 1949 and operated until 2010. (Courtesy of the Mount Vernon Mayor's Office.)

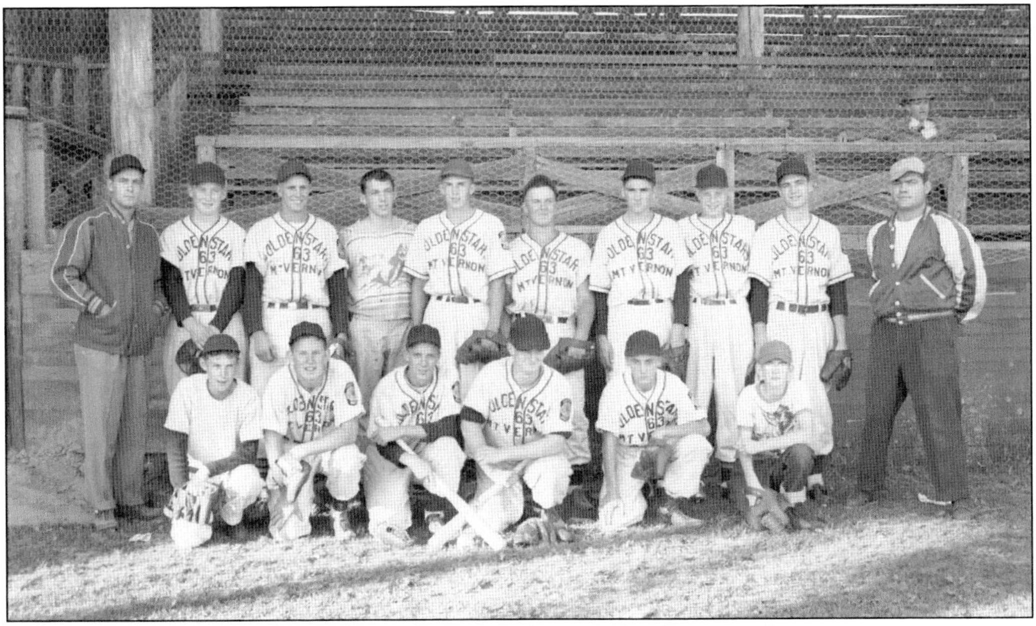

The Mount Vernon Golden Stars, seen here in 1950, were the American Legion team representing Mount Vernon's Post 63. Founded in 1925, American Legion Baseball was the first program in the world to provide a national baseball tournament for teenagers. (Photograph by Bill Forman; courtesy of Mark Iverson.)

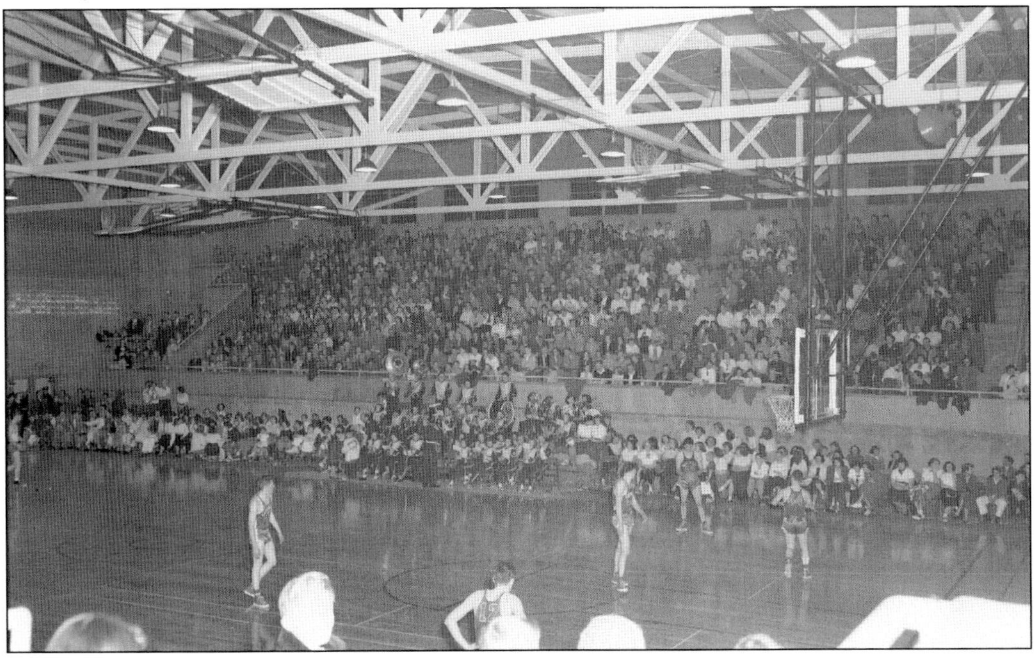

Over 2,000 spectators crammed into the high school's new gym on the evening of January 26, 1951, for opening night festivities. The new facility replaced the original 1922 gymnasium. The crowd watched the Mount Vernon Bulldogs lose to the Nooksack Valley Pioneers 47-42, and then the Skagit Valley Junior College Cardinals lost to the University of Washington freshman team 50-40. (Photograph by Bill Forman; courtesy of Mark Iverson.)

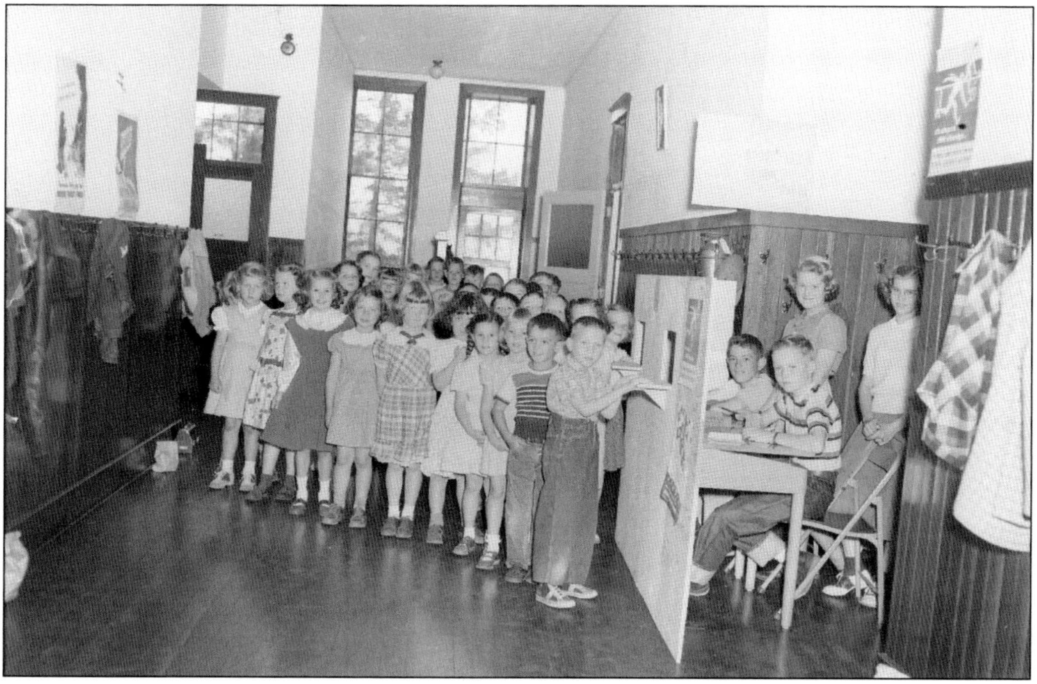

A group of unidentified students at Roosevelt Middle School line up at the school's student-run bank in 1951. The bank was sponsored by First National Bank, which had a long history in Mount Vernon. When First National was organized in 1891, it was the only bank in the city, with a capital stock of $25,000. (Photograph by Bill Forman; courtesy of Mark Iverson.)

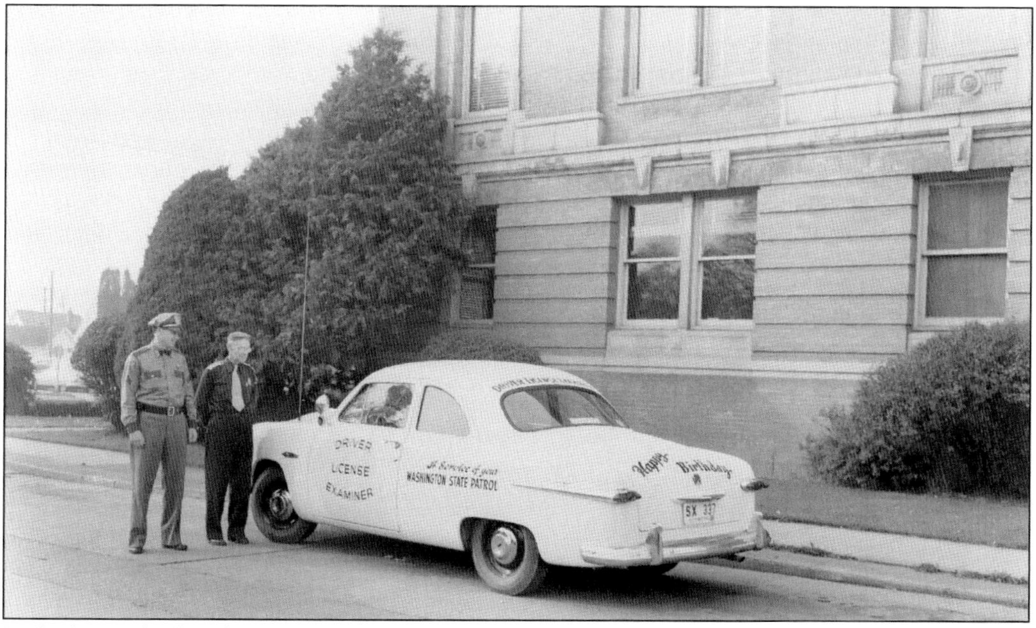

Parked outside the Skagit County Courthouse on November 9, 1952, is the newest Washington State Patrol car in Mount Vernon. The car was designated as the "driver license examiner" and displayed birthday wishes for those taking the test. (Photograph by Bill Forman; courtesy of Mark Iverson.)

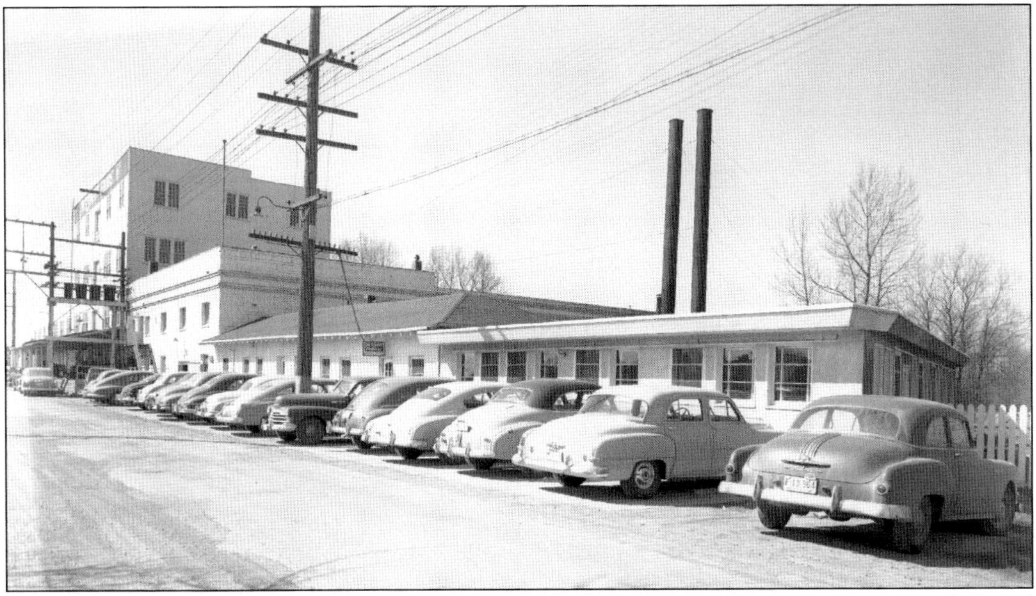

Above is a photograph of the PictSweet processing plant, taken in April 1951. PictSweet Foods, based in Mount Vernon, began as a one-plant, private-label canning operation and eventually became one of the top frozen food packing companies in the country. PictSweet was a major employer in the area, both for factory workers and for the farmers who supplied the plant. Below is a scene from an open house at the processing plant in June 1953. By the 1950s, PictSweet had plants and warehouses in Washington, Oregon, California, and Florida. In July 1954, the organization was purchased by an Indianapolis-based company named Stokely-Van Camp, Inc. The headquarters for the new Stokely-PictSweet frozen foods division remained in Mount Vernon until it moved to Oakland, California, in 1957. (Both, photograph by Bill Forman; courtesy of Mark Iverson.)

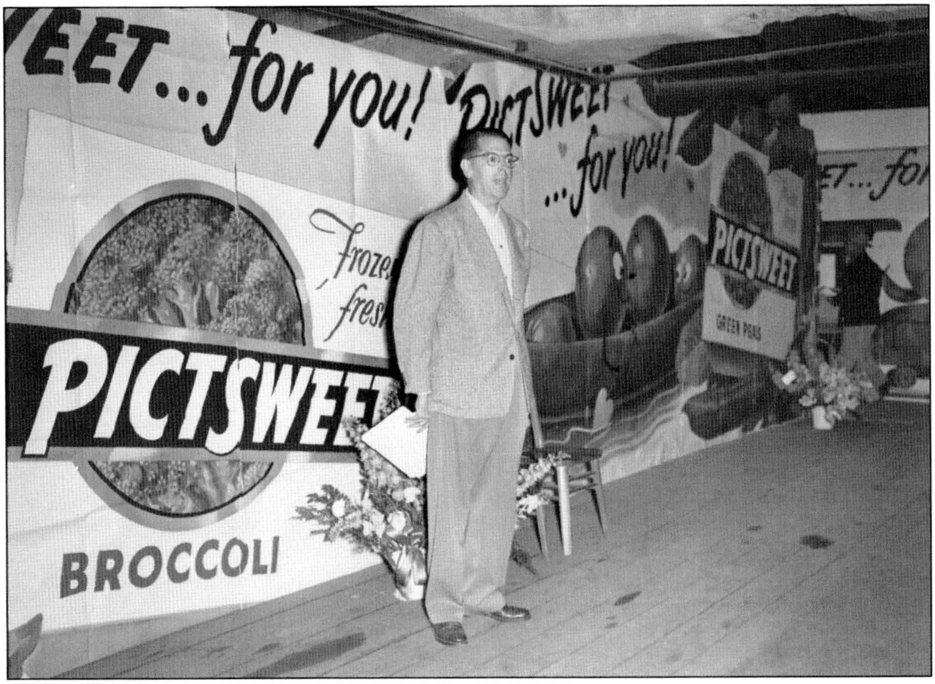

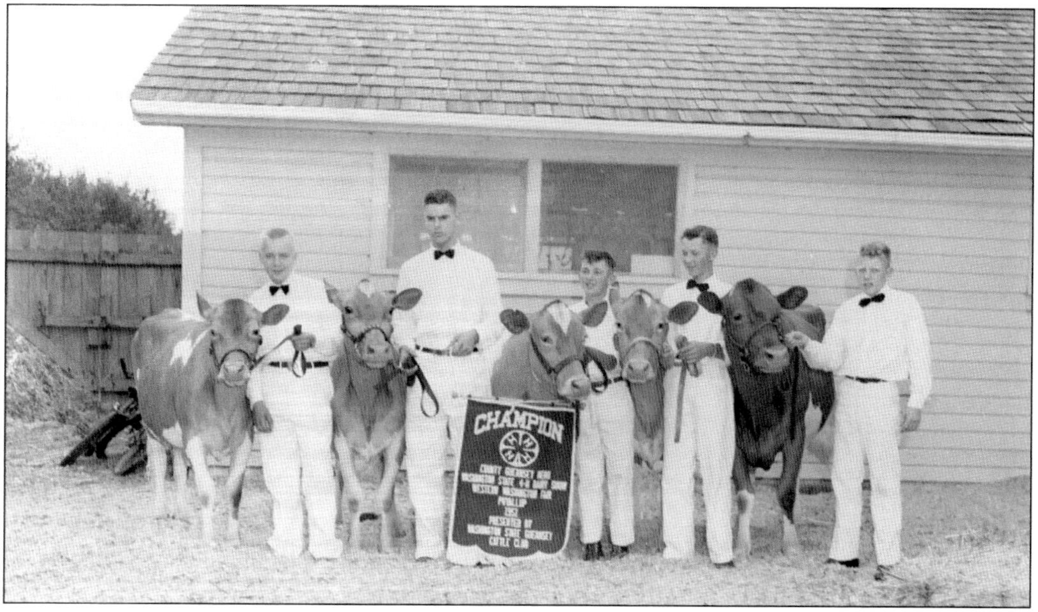

Started in 1902, 4-H is considered one of the first youth development organizations in the United States. Since the 1920s, 4-H has been active in Mount Vernon, revolutionizing how youth connect to practical, hands-on learning experiences while outside the classroom. Above are members of Mount Vernon's 4-H chapter during the 1951 fair. Seen below are their counterparts, members of Mount Vernon's Future Farmers of America (FFA), at the same fair. Founded in 1928, the FFA brought together students, teachers, and agribusiness to solidify support for agricultural education. The Mount Vernon FFA Chapter, established in 1932, is one of the original charter FFA chapters in the State of Washington. Every summer, members of both organizations actively participate in Mount Vernon's Skagit County Fair. (Both, photograph by Bill Forman; courtesy of Mark Iverson.)

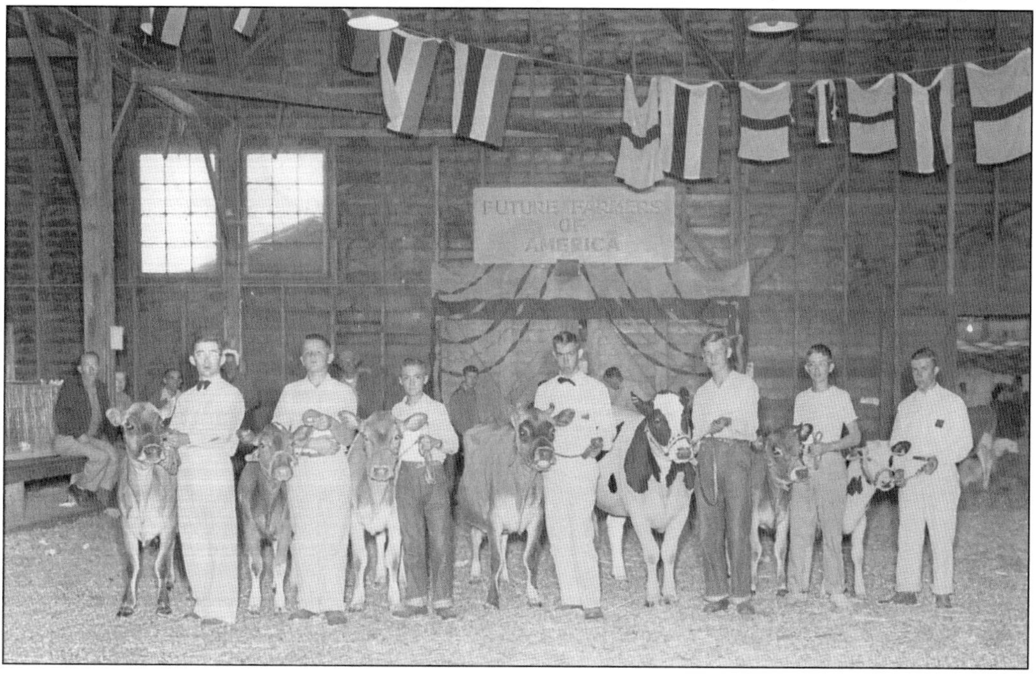

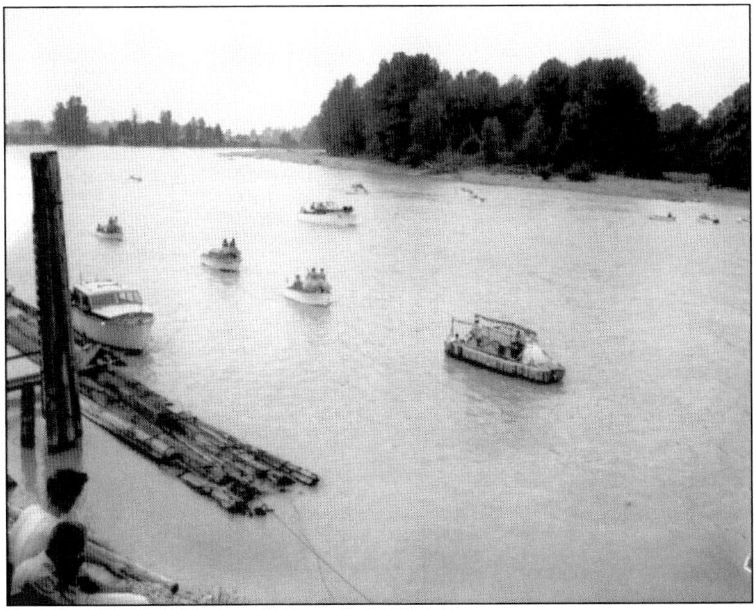

During the August 1952 Skagit County Fair (also called the 4-H Fair), a parade on the Skagit River featured Queen of the Fair Rosemary Anderson and her court. They floated down the river in several small boats while a crowd watched from the revetment. At that time, the river was still embraced as a focal point of downtown. (Photograph by Bill Forman; courtesy of Mark Iverson.)

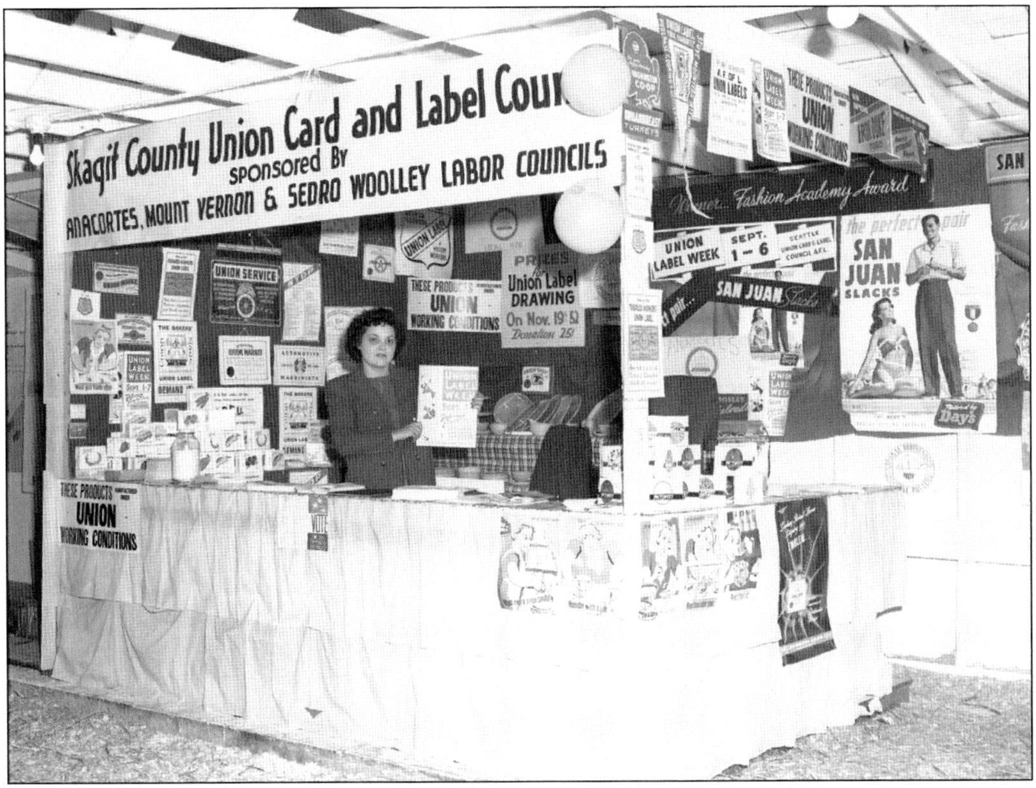

Seen here at the 4-H Fair in 1952 is a booth sponsored by Union Labor Councils. It is interesting to note some of the products featured in the picture. One sign promotes Union Label Week, a time traditionally set aside for consumers to make a special effort to support jobs by looking for union-made goods and union-produced services when they shop. (Photograph by Bill Forman; courtesy of Mark Iverson.)

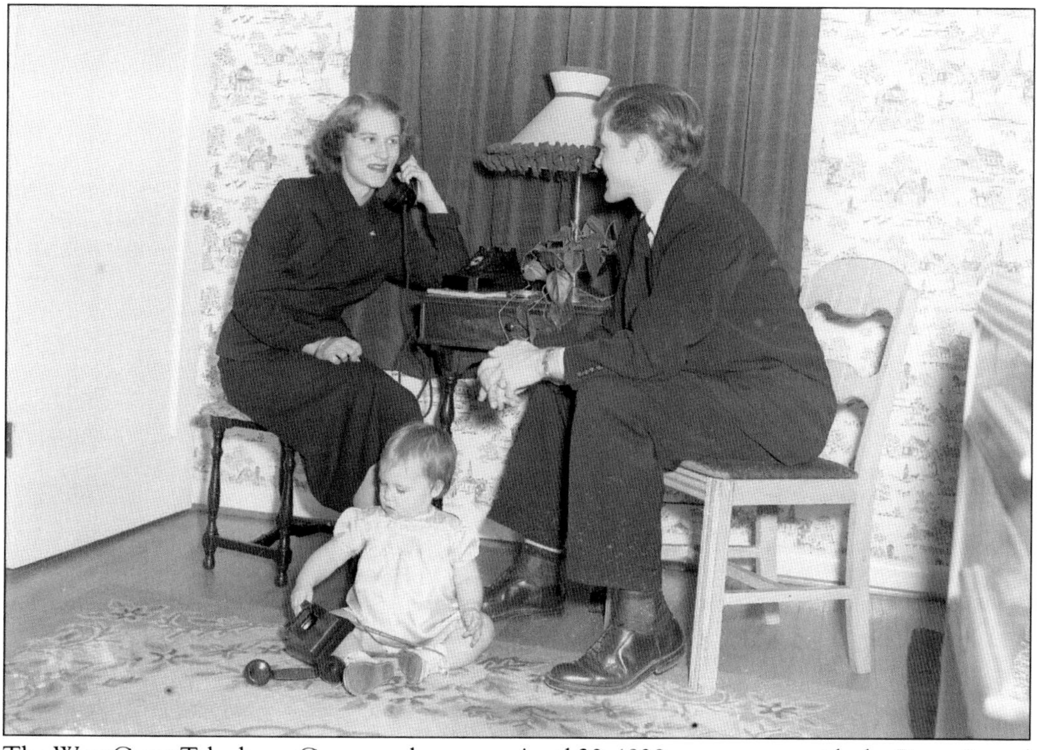

The West Coast Telephone Company began on April 30, 1928, as a merger with the Puget Sound Telephone Company and several smaller companies in Oregon and California. At that time, the company had 34,895 telephones in service. In October 1952, the company installed its 100,000th telephone in Mount Vernon. The telephone was installed by Keith Breathour in the residence of Mr. and Mrs. Don Mahaffey (seen above with daughter Marsha Anne) at 519 Washington Street. The Mahaffeys were then given a tour of the "telephone plant," seen below, before being treated to a celebratory dinner at the President Hotel with company representatives. In 1964, West Coast Telephone Company would merge with General Telephone Company of the Northwest to form GTE Northwest. (Both, photograph by Bill Forman; courtesy of Mark Iverson.)

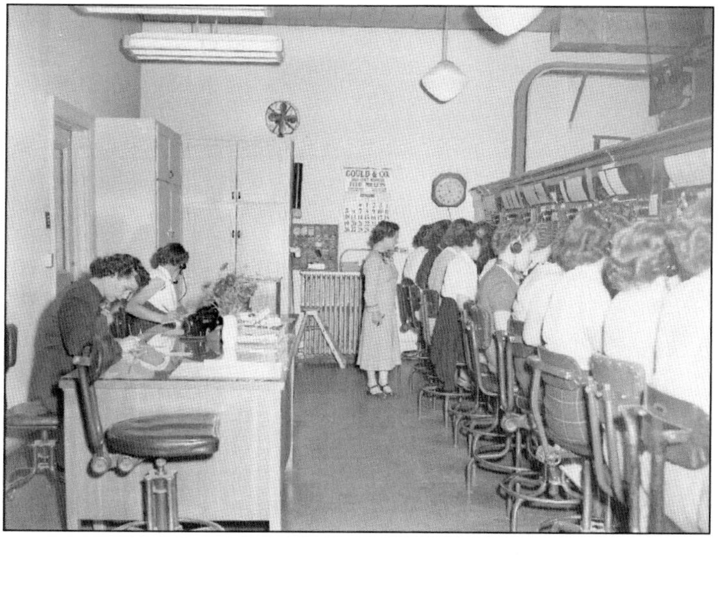

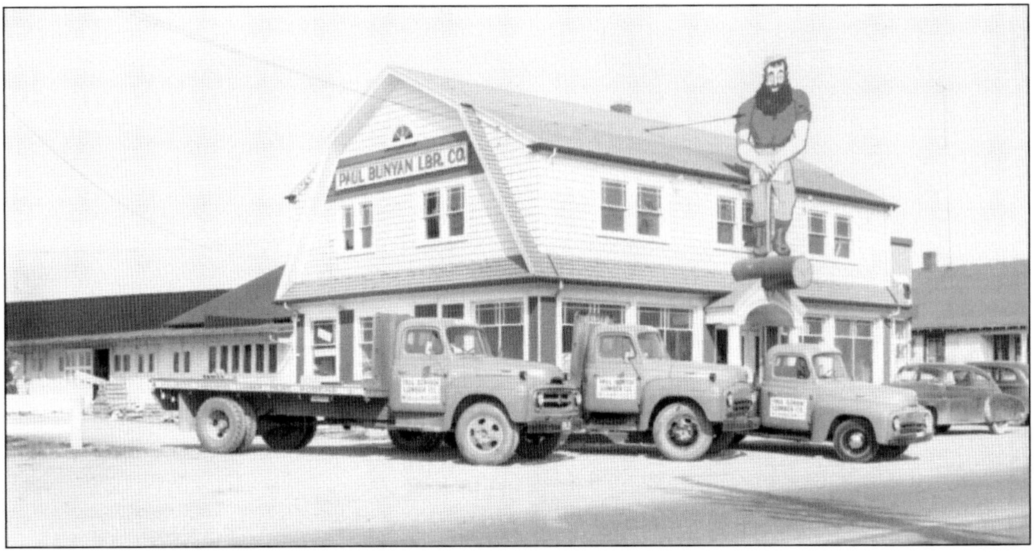

In 1952, John D. Connell founded the Paul Bunyan Lumber Company in Mount Vernon. He later opened a second store in Oak Harbor. Connell was very active in youth activities. He coached one of the first Little League teams in Mount Vernon as well as established the Youth Recreation Program at Hillcrest Park. (Photograph by Bill Forman; courtesy of Mark Iverson.)

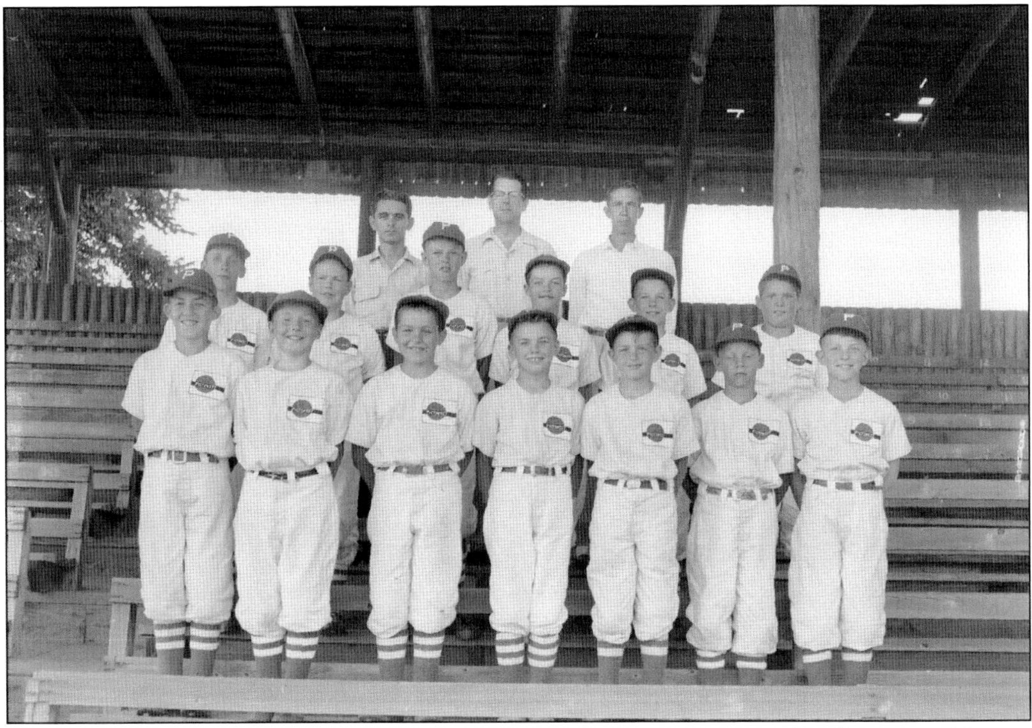

Founded by Carl Stotz in 1939, Little League came to Skagit County in 1953. The Mount Vernon program, headed by James H. Hiester, began with four teams sponsored by Kiwanis, Rotary, Lions, and PictSweet. The first year, 80 boys between the ages of 10 and 12 participated. Pictured here on July 13, 1954, is the PictSweet Little League Team (players unidentified). (Photograph by Bill Forman; courtesy of Mark Iverson.)

For many years, the area known as "the Hill," east of downtown, has been a desirable location for homebuilders. It is interesting to note the prices of new homes built there in 1954. In 2012, those same homes are valued at approximately 10 times their original sales prices. (Photograph by Bill Forman; courtesy of Mark Iverson.)

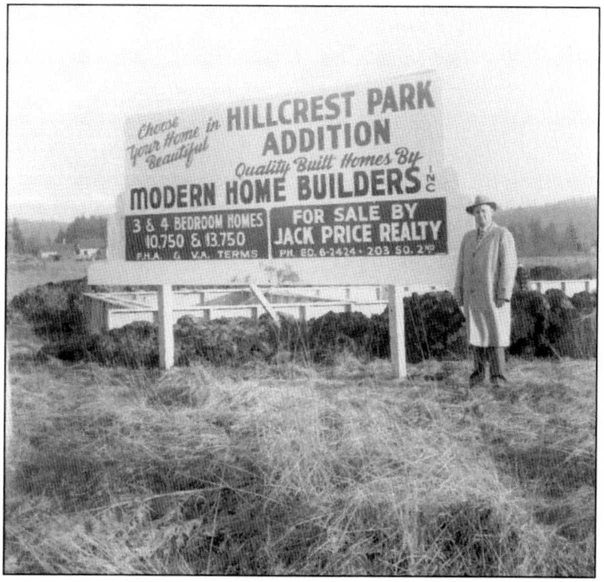

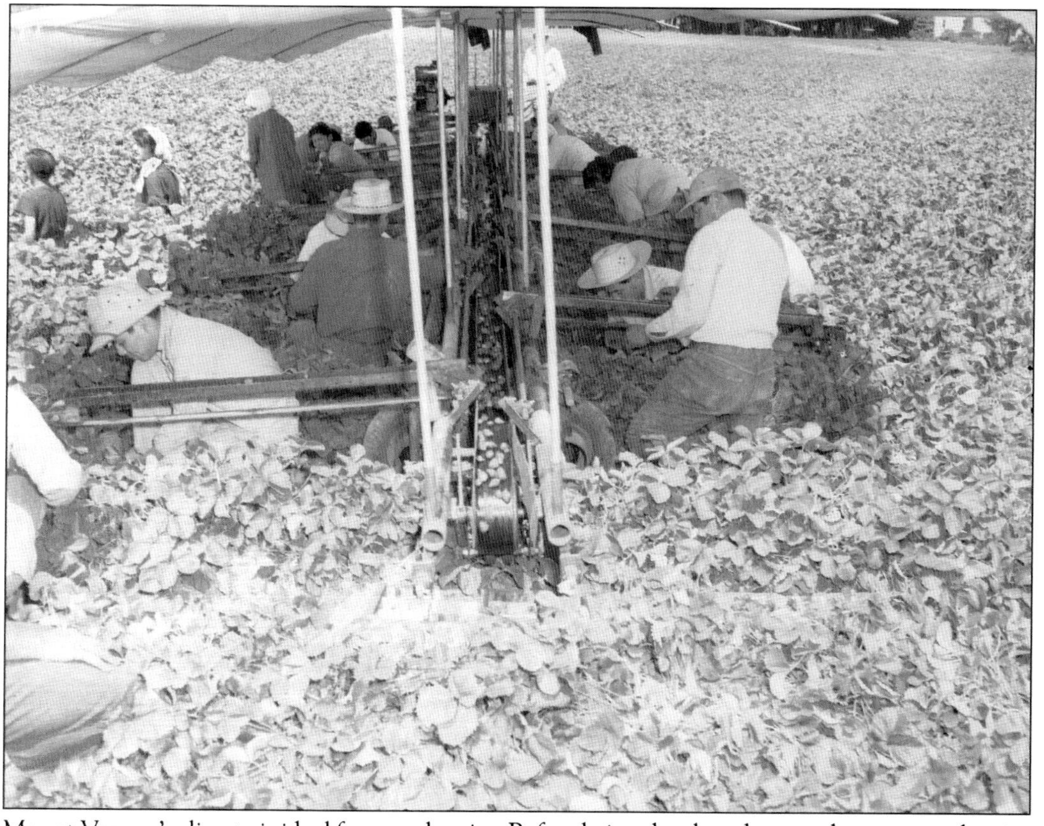

Mount Vernon's climate is ideal for most berries. Before being developed as a cash crop, strawberries grew easily around the many stumps on land that had been logged. Photographed in July 1953, these strawberry pickers are using a "berry belt." Varieties included Northwest and Puget, but the Shuksan, developed in 1970 by WSU researchers, became the standard due to its hardiness. (Photograph by Bill Forman; courtesy of Mark Iverson.)

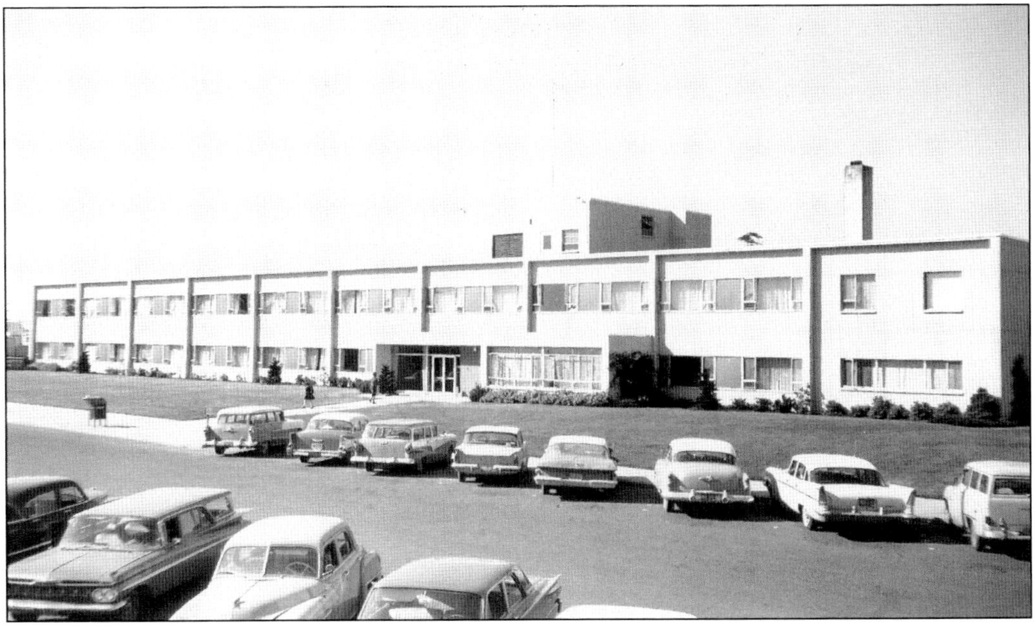

By May 1954, the Women's Guild of Skagit General Hospital (established in 1949) had raised more than $10,000, which they used to purchase land (between Fourteenth and Fifteenth Streets and Kincaid and Montgomery Streets) to be the future site of a community-owned hospital. In September 1955, with a vote of 2,529 to 217, voters passed a $550,000 bond issue to fund the new facility. In April 1956, passage of a second bond issue for $120,000 allowed plans to be adjusted from the originally proposed 65-bed hospital to a 78-bed hospital. The ground-breaking ceremony was held on January 3, 1957, and on April 9, 1958, the hospital (above) officially opened to patients. Eric Lefstad, son of Nancy and Dana Lefstad, was the first baby born at the new state-of-the-art facility, just three hours after its opening. (Both, courtesy of Skagit Valley Hospital.)

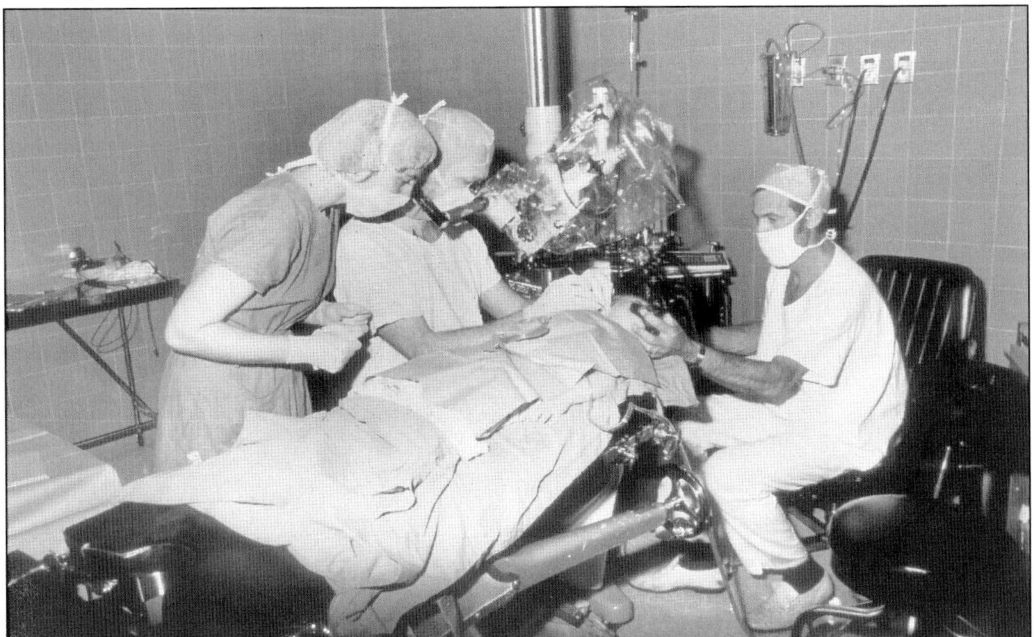

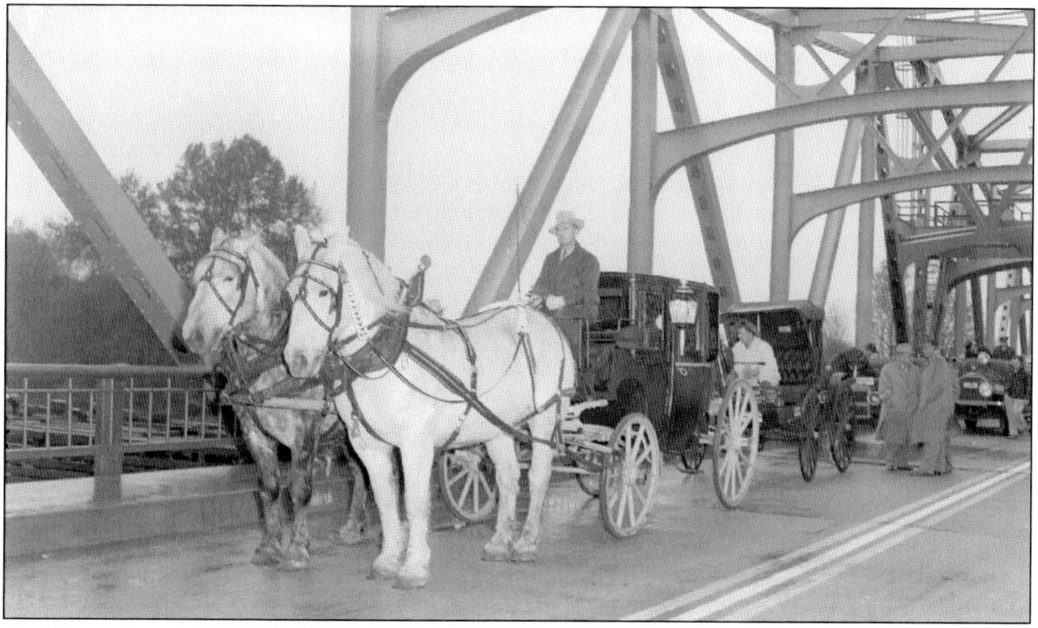

On November 4, 1954, the West Side Bridge (spanning the Skagit River at downtown Mount Vernon) opened to great fanfare. It was built to replace the former 1908 bridge. Following the official ribbon cutting by Gov. Arthur B. Langlie, approximately 12,000 cars streamed over the newly completed, $1.25-million bridge. The coach in the photograph above was lent to the parade by Robert Tarry and carried Governor Langlie across the bridge, first in the procession. Below is a 1902 Holsman, one of six "ancient" cars and two horse-drawn vehicles that participated in the bridge-opening parade. In 2007, average daily traffic on the bridge was over 21,000. (Both, photograph by Bill Forman; courtesy of Mark Iverson.)

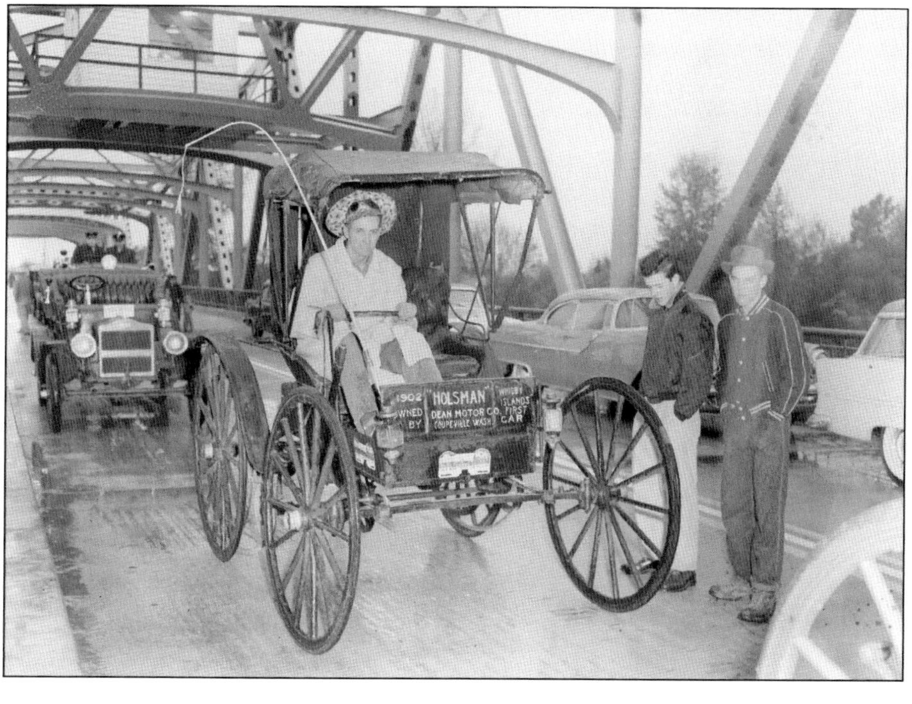

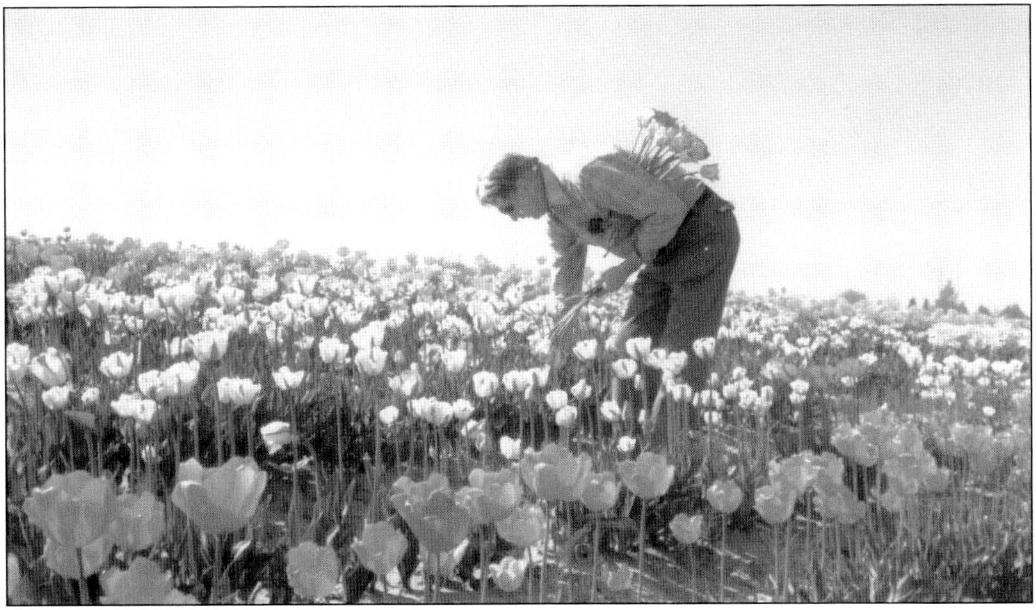

William Roozen was born in Holland to a family of prominent bulb growers that had been involved in the industry since the mid-1700s. In 1947, he and his wife, Helen, came to Mount Vernon. They purchased a partnership in the Washington Bulb Company in 1955 and later assumed full ownership. With the help of their 10 children, the Roozens' company became the largest tulip-bulb grower in the country. (Courtesy of Lisette Mast.)

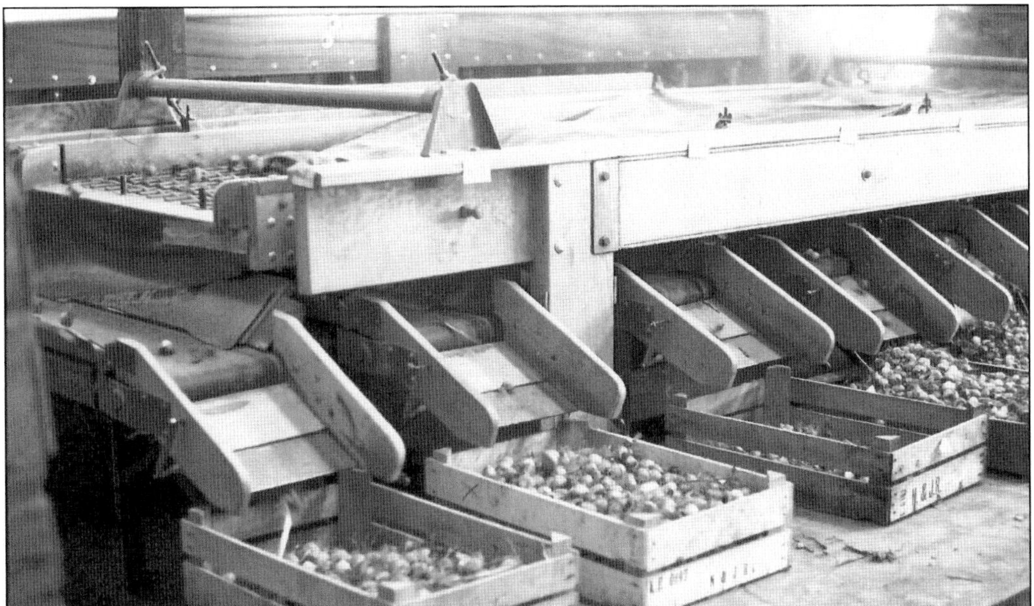

Tulip bulbs were harvested by hand before being run through a bulb sorter like the one seen here. Large bulbs were sold, and smaller "babies" were replanted to create new families of bulbs. Shortly after blooming, tulips are "topped" (the blooms are cut off) so that nutrition goes into bulb growth rather than into the flower. Keeping the bulb varieties separated is crucial. (Courtesy of WSU Skagit County Extension.)

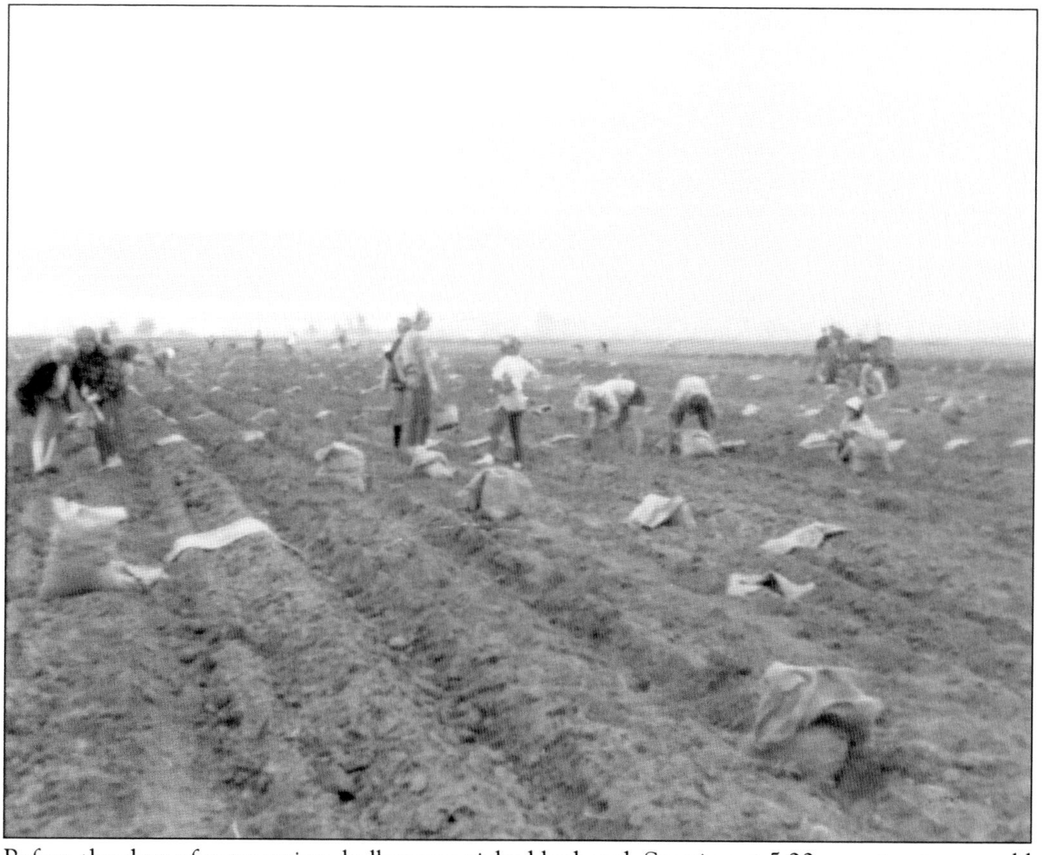

Before the days of automation, bulbs were picked by hand. Starting at 5:00 a.m., tractors would turn up the dirt before the workers began pulling out and sacking the bulbs, being careful to not mix varieties. Rows were marked off in 200-foot sections, and pickers would work through the field one section at a time. When a section was complete, a field boss would come check the work and punch the worker's card, like the one seen below (pickers were paid by the foot). Picking was a popular summer job for children, and a large field could have up to 250 workers while three to five field bosses supervised. (Both, courtesy of Lisette Mast.)

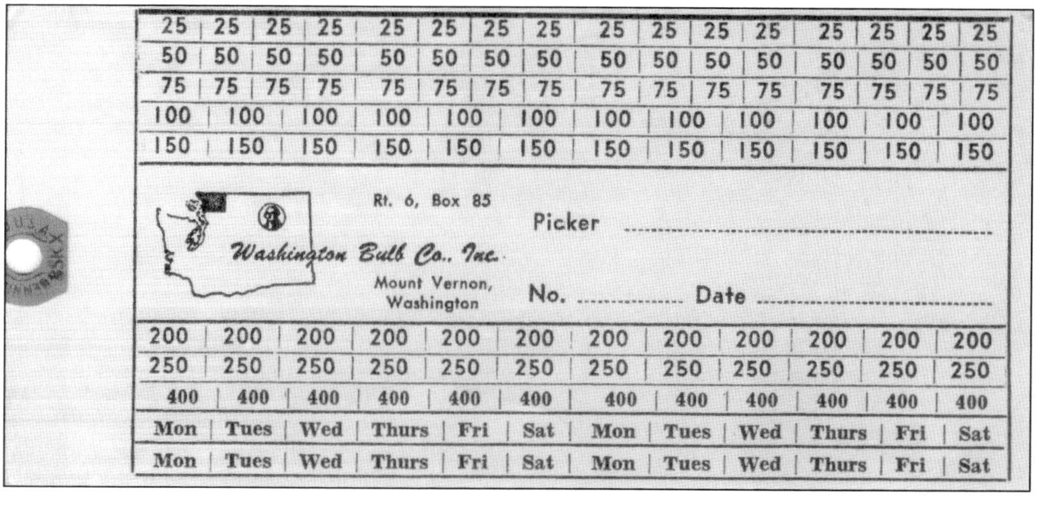

25	25	25	25	25	25	25	25	25	25	25	25	25	25	25	25
50	50	50	50	50	50	50	50	50	50	50	50	50	50	50	50
75	75	75	75	75	75	75	75	75	75	75	75	75	75	75	75

100	100	100	100	100	100	100	100	100	100	100	100
150	150	150	150	150	150	150	150	150	150	150	150

Rt. 6, Box 85
Washington Bulb Co., Inc.
Mount Vernon, Washington

Picker ....

No. .... Date ....

200	200	200	200	200	200	200	200	200	200	200	200
250	250	250	250	250	250	250	250	250	250	250	250
400	400	400	400	400	400	400	400	400	400	400	400
Mon	Tues	Wed	Thurs	Fri	Sat	Mon	Tues	Wed	Thurs	Fri	Sat
Mon	Tues	Wed	Thurs	Fri	Sat	Mon	Tues	Wed	Thurs	Fri	Sat

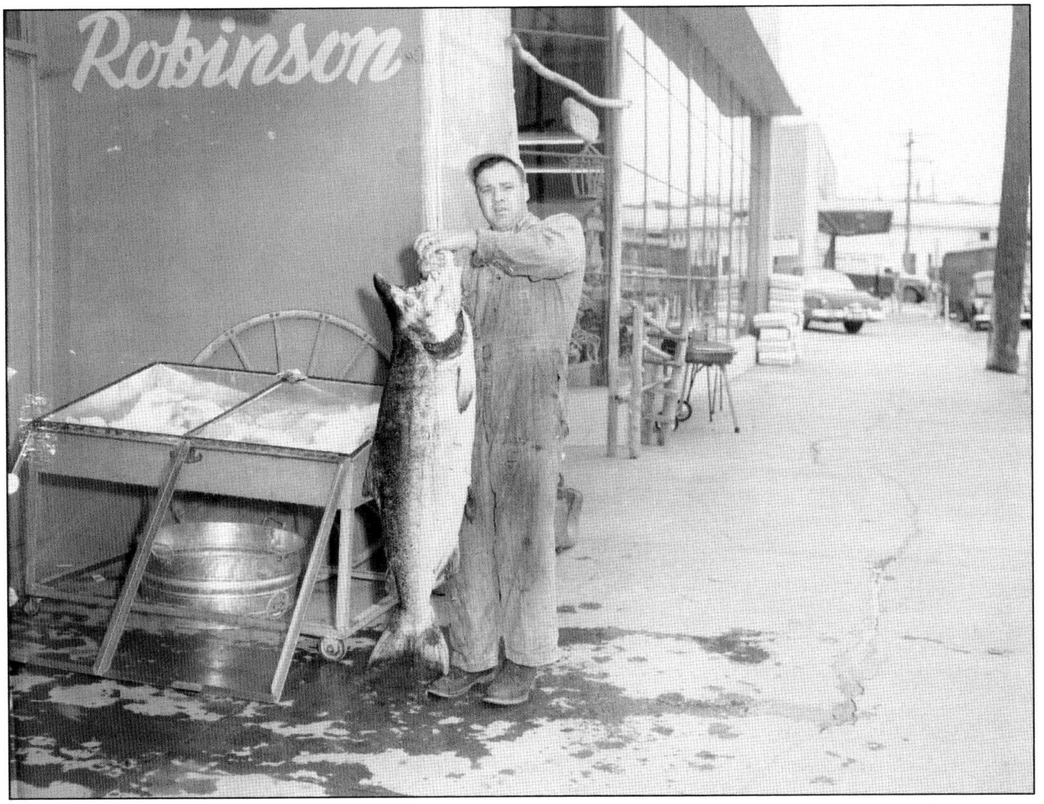

The Skagit River provides an important spawning habitat for salmon, being the only large river system in Washington that contains healthy populations of all five native salmon species (chinook, coho, chum, pink, and sockeye) and two species of trout (steelhead and cutthroat). All can be fished during their seasons. This unidentified fisherman had a lucky day in July 1955. (Photograph by Bill Forman; courtesy of Mark Iverson.)

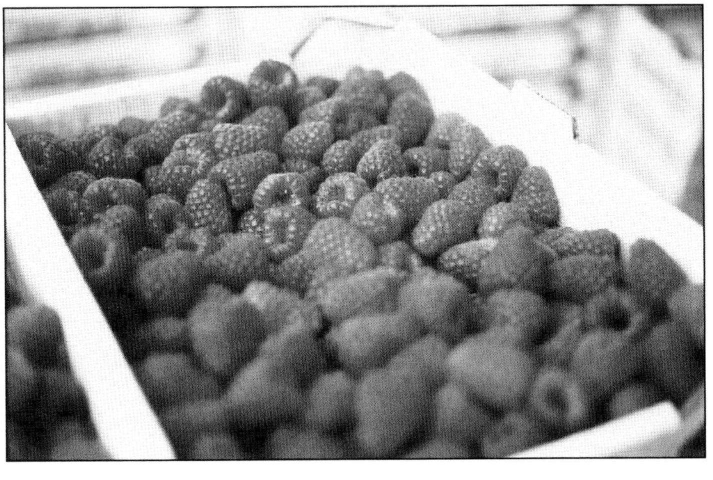

Raspberries have been an important and popular crop in the area since the 1960s. The main cultivar since 1967 has been the Meeker variety, developed by a WSU agent in Puyallup, Washington. It is a vigorous plant, with long fruiting laterals. Advances in machinery have allowed this berry to be machine-harvested since the late 1980s. (Courtesy of WSU Skagit County Extension.)

Herman I. "Herm" Hanson was the mayor of Mount Vernon from 1962 until 1971. He also spent several years as the PUD commissioner. One of the greatest accomplishments of his administration was working with the county and state to obtain funds in order to fix what had been a traffic mess at the end of First Street. (Courtesy of the Mount Vernon Mayor's Office.)

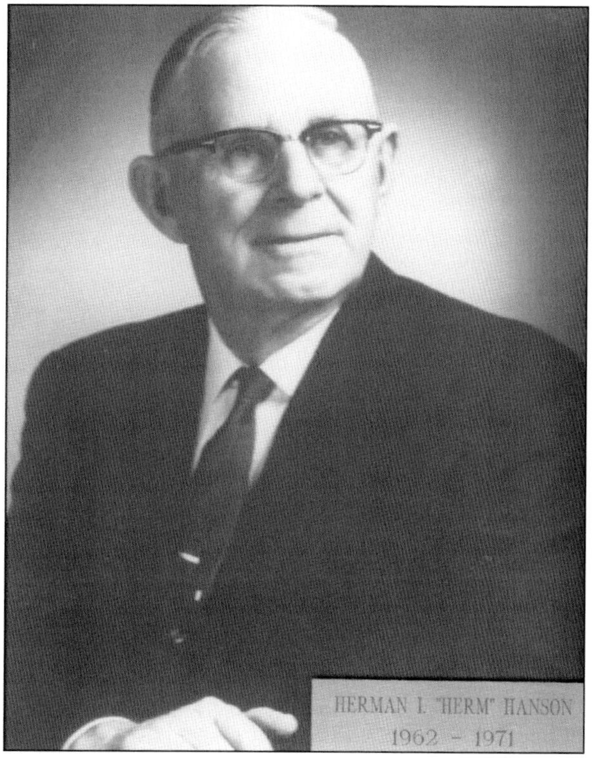

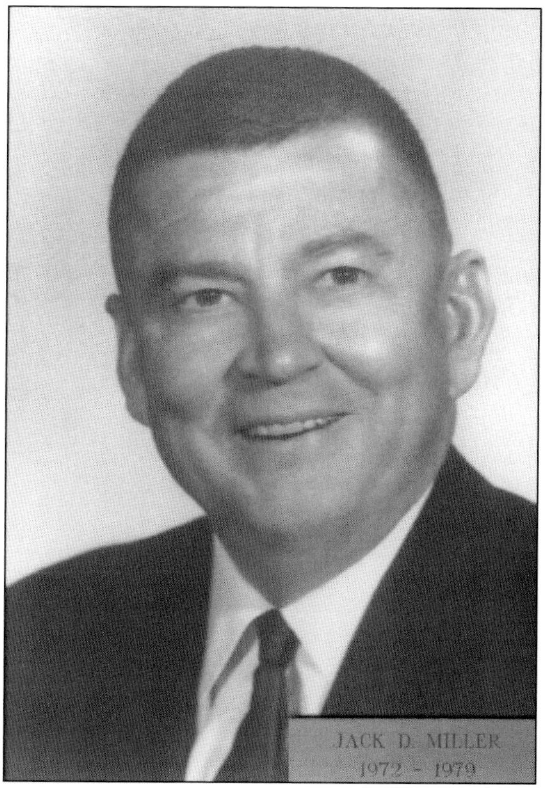

Jack Miller served as mayor from 1972 until 1979. During his time in office, he acquired $500,000 in federal funds for a massive downtown improvement project. A sewer line was run down the entire length of First Street, which allowed for the complete reconstruction of the antiquated street. The project also turned part of Pine Street into a paved courtyard. (Courtesy of the Mount Vernon Mayor's Office.)

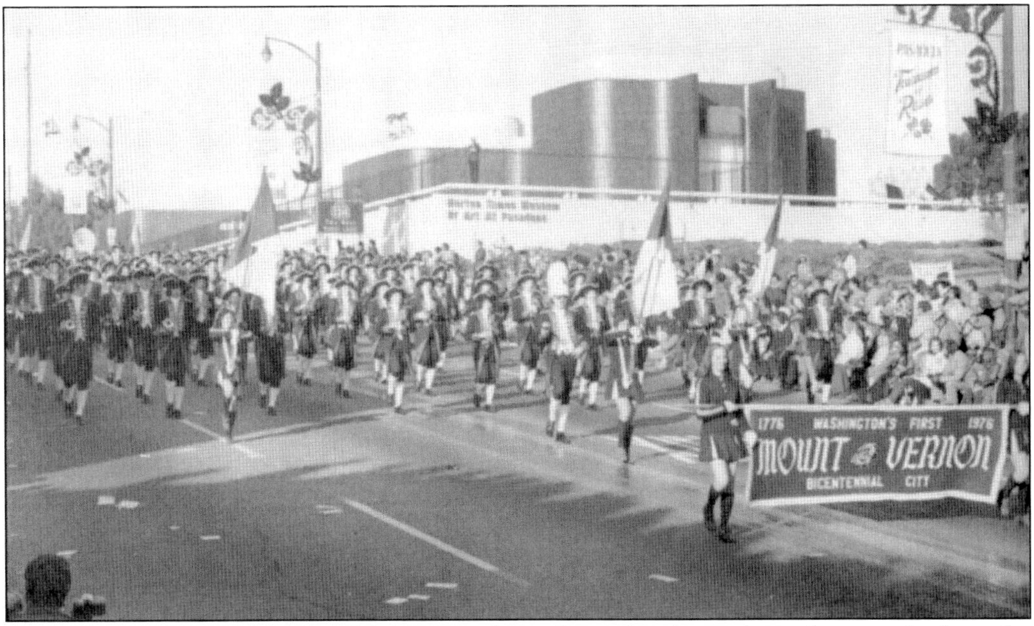

In 1973, William Beck, owner of the City Bakery, devised a plan to bring business back to the struggling downtown area. He wanted the town to pay homage to American history and evoke a Colonial style. Beck promoted his plan with a patriotic group, the Mount Vernon Volunteer Militia, who paraded in hand-sewn reproduction Colonial uniforms. Although ultimately unsuccessful, the plan was supported by some in the community, most notably the high school. In 1974, they purchased Colonial-style band uniforms, and on January 1, 1976, the high school band (above) and the militia (below) were part of the Tournament of Roses Parade in Pasadena. While in Southern California, the band also marched in a Disneyland parade on December 29, 1975. (Both, photograph by John Sobieski; courtesy of the Mount Vernon School District.)

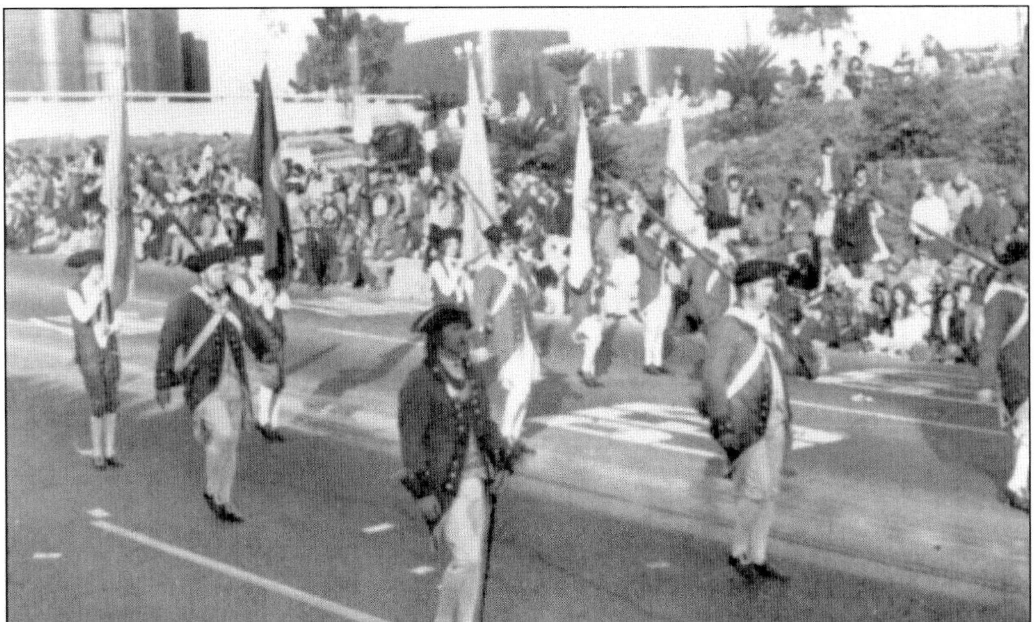

Eight

Revitalization and Reinvention 1980–2012

In 1984, the chamber of commerce inaugurated the Skagit Valley Tulip Festival. Events were planned to enhance the experience of tourists coming to view the spring daffodils and tulips. In 1994, the Tulip Festival broke off from the chamber and became an entity of its own. Originally lasting three days, the festival has grown over the years to span the entire month of April.

The cattle and dairy industry has continued to thrive, but with the closing of the creameries, milk had to be transported down to Seattle for processing. Over the years, canning and processing plants also left the area. Agriculturally, berries, peas, potatoes, apples, and seeds continue to expand in production. The WSU Mount Vernon Northwestern Washington Research and Extension Center has been instrumental in advancing these crops technologically.

The citizens of the community remain committed to improvement, as evidenced by the continual support given to school levies. The increase in population and financial support has allowed for retrofitting and additions to Lincoln, Washington, Madison, Union High School, and Jefferson as well as construction of Centennial (1989), Little Mountain (1996), and Mount Baker Middle School (1997).

There is a strong push to rejuvenate Mount Vernon's historical downtown. The Mount Vernon Downtown Association joined the Main Street Program, administered through the National Trust for Historic Preservation, in 2008. However, any revitalization will be in vain if the constant threat of flooding and the prohibitively high cost of flood insurance are not addressed. The Downtown and Waterfront Area Master Plan was completed in 2008. At an estimated cost of approximately $38 million, its key objective is to provide flood protection. The overall three-phase project, slated to finish in 2013, will entail 9,300 feet of flood wall and levy, 4,200 feet of river walk and trail connection, and 64,000 square feet of public riverfront plaza.

Clearly, the tiny riverfront outpost has grown over the years. Today, Mount Vernon, gateway to the San Juan Islands and Mount Baker, has a population of over 32,000 and is a thriving city ready for the challenges and changes in the years ahead.

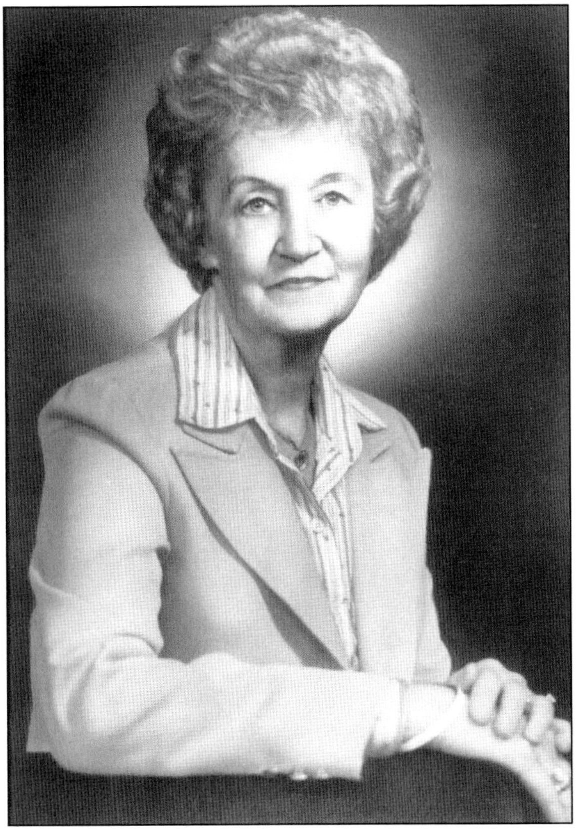

Ruth E. Gidlund, born February 12, 1914, served as the first female mayor of Mount Vernon from 1980 to 1983. The dominating issue of her term was the management of the Skagit River flood plan. Strict construction regulations (based on high 100-year flood levels) were proposed by the Federal Emergency Management Agency (FEMA); enforcement of those regulations would impede development. The levels were eventually reassessed and decreased. (Courtesy of Kay Dewey.)

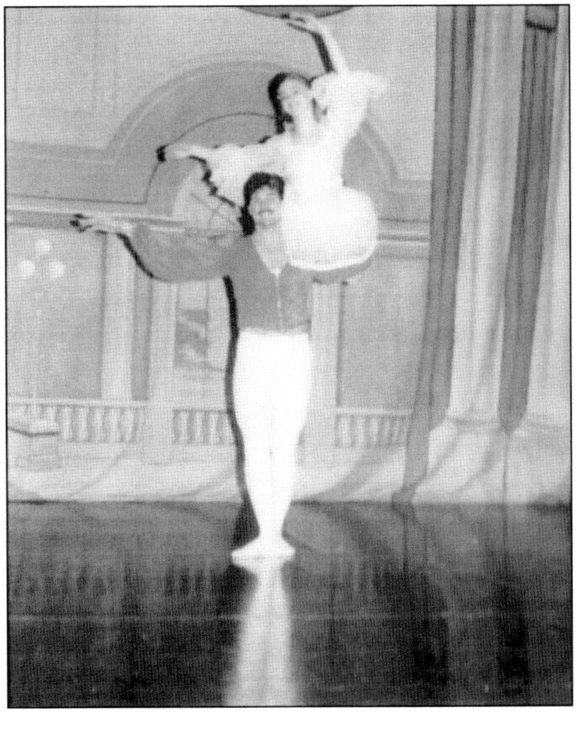

Since 1990, the Skagit Valley Academy of Dance has held an annual performance of *The Nutcracker*. Pictured here are the first Clara (Renee Kinnear) and Prince (Ralph Fandel). (Courtesy of Skagit Valley Academy of Dance.)

Skye K. Richendrfer was mayor from 1996 until 2003. Major accomplishments during his time in office included the completion two new fire stations, the police/courts campus, and numerous road and transportation projects (including the Burlington-Mount Vernon Bridge at Riverside and Skagit Station) as well as fiber-optic network deployment and the city's purchase of the Lincoln Theatre. A talented bagpiper, Richendrfer also founded the Skagit Valley Highland Games and Celtic Arts Foundation. (Courtesy of the Mount Vernon Mayor's Office.)

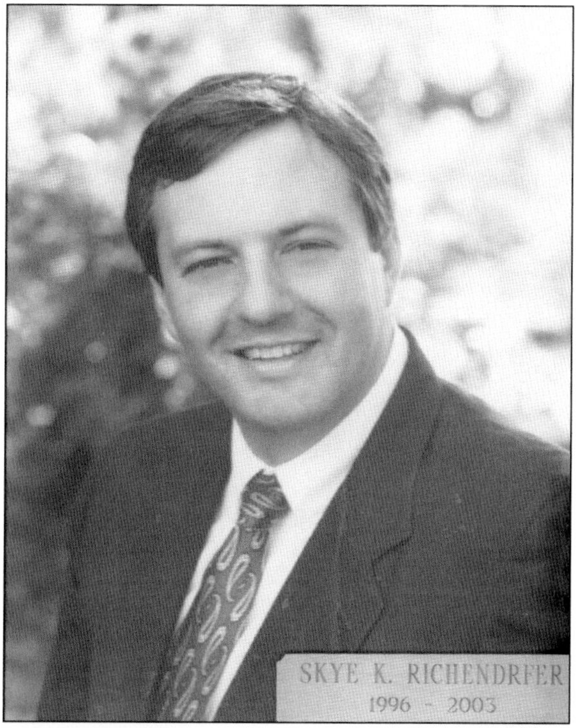

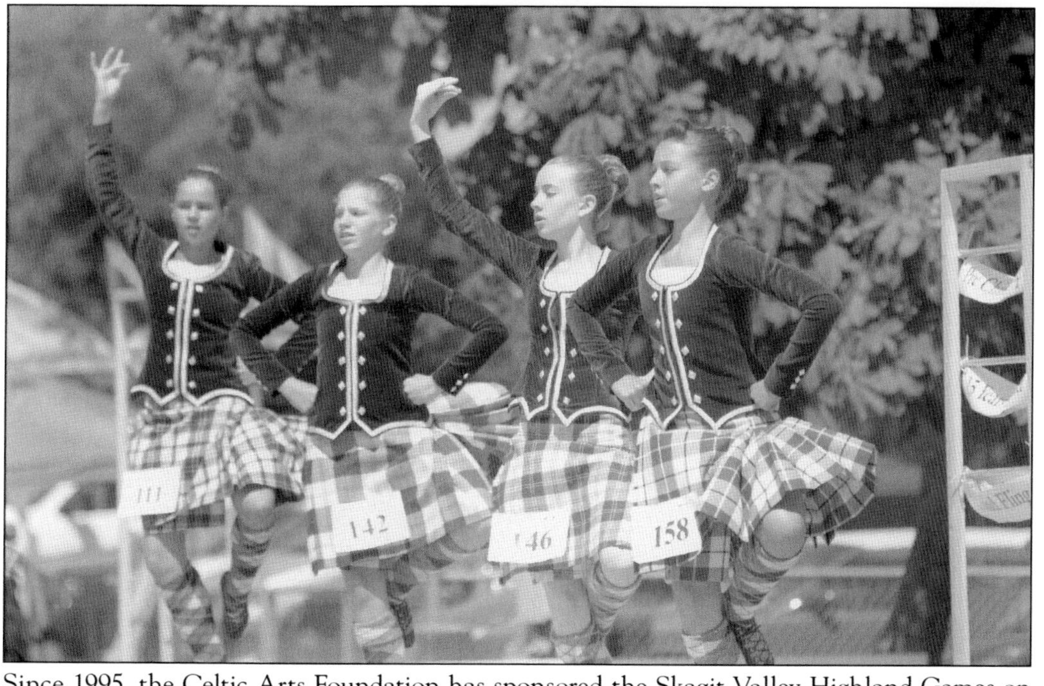

Since 1995, the Celtic Arts Foundation has sponsored the Skagit Valley Highland Games on the second full weekend in July. The festival is part of a yearly circuit of competitions held in the Pacific Northwest and the Lower Mainland of British Columbia; therefore, the quality of performers and competitors is extremely high. These unidentified girls are competing in highland dancing. (Courtesy of the Celtic Arts Foundation.)

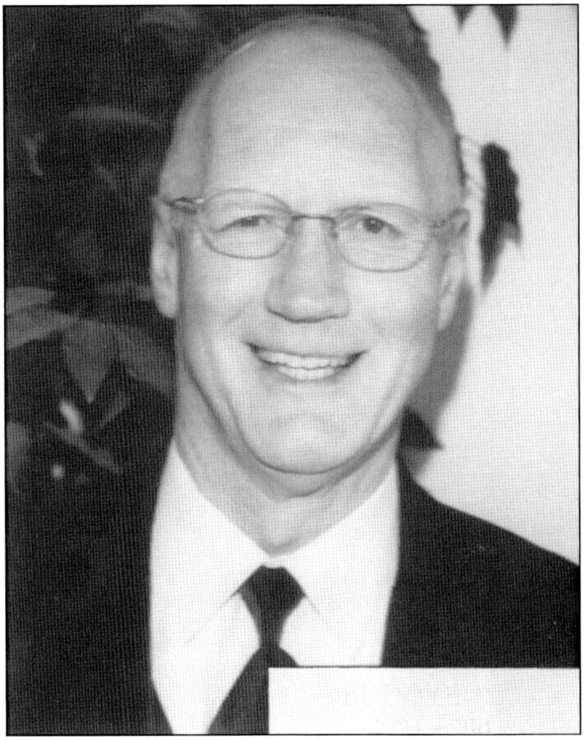

Bud Norris served as mayor of Mount Vernon from 2004 until 2011. His lasting legacy is the (still-in-progress) Downtown and Waterfront Area Master Plan. Slated to be complete in 2013, the massive construction project should remove downtown from the 100-year flood map and revitalize the area. Norris was also able to maintain a budgetary surplus through difficult economic times. (Courtesy of the Mount Vernon Mayor's Office.)

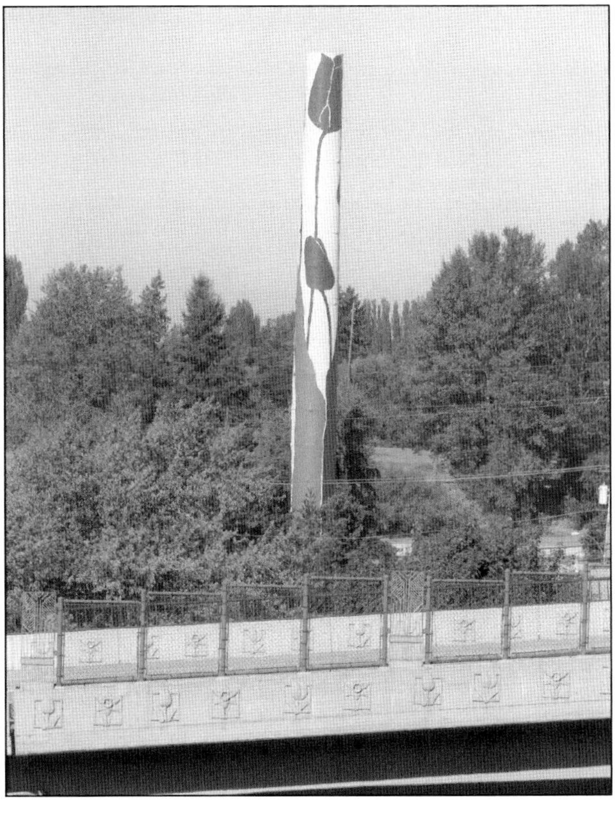

Rising above downtown, the tulip smokestack is a remnant of the once-thriving Carnation Company condensery. In 2007, the top section of the stack began falling, creating a safety hazard. Instead of tearing it down, the community rallied to have the stack repaired and painted with brightly colored tulips. The stack is now an iconic symbol of Mount Vernon. (Courtesy of Jessica Bylund.)

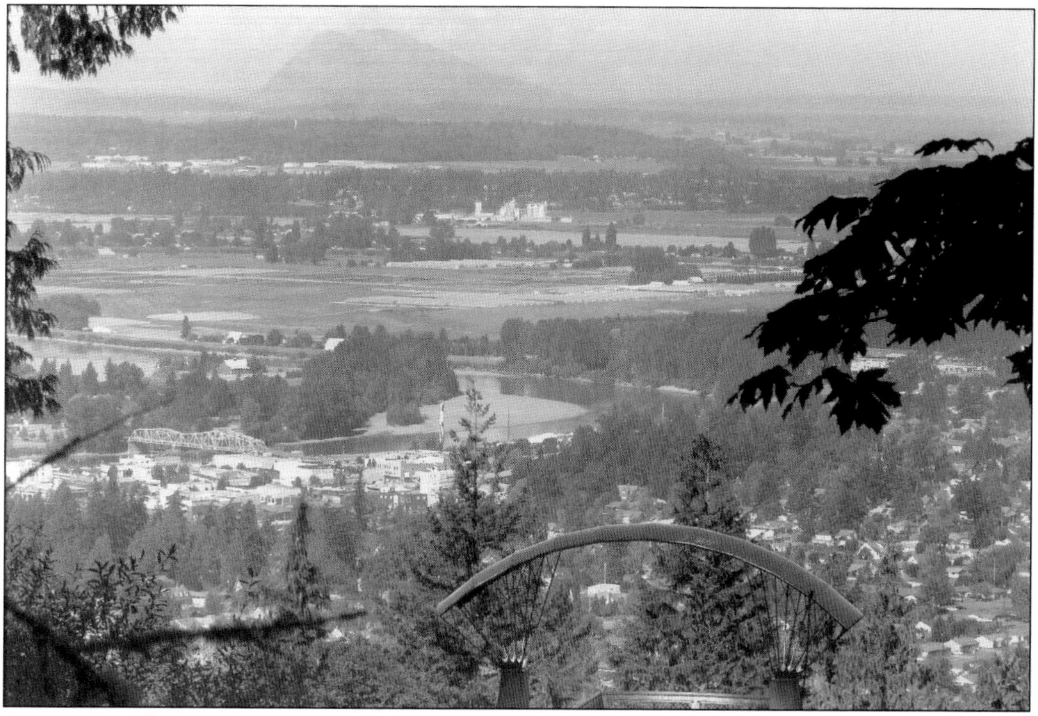

Originally called English Park because of its origin as a 240-acre donation by the English Logging Company, Little Mountain Park now encompasses 522 acres of forest and, with an elevation of 934 feet, is the highest point in Mount Vernon. It currently has almost eight miles of trails, and in 2005, the north viewpoint (pictured) was opened, granting spectacular views of downtown, Skagit River, Lummi Island, and Mount Baker. (Courtesy of Jessica Bylund.)

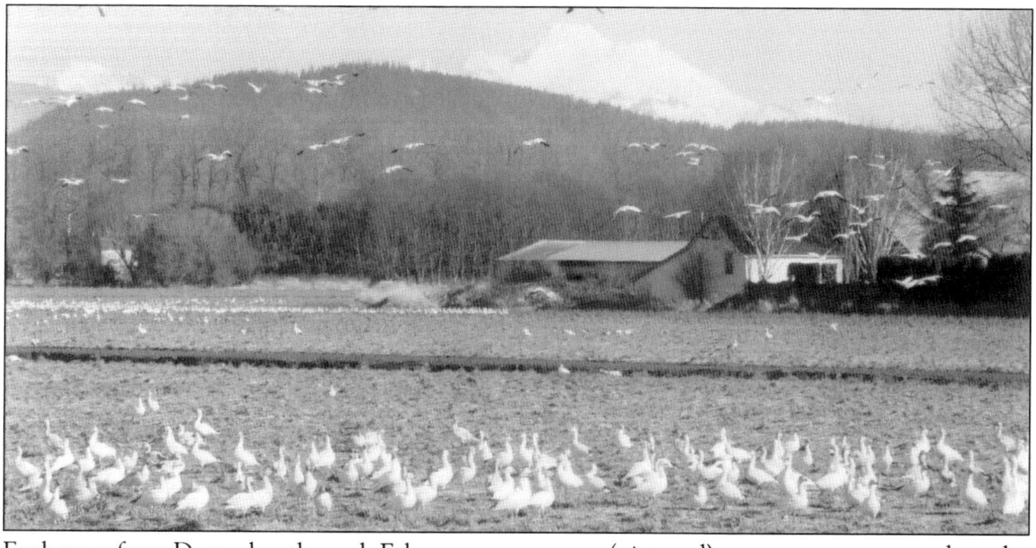

Each year, from December through February, snow geese (pictured), trumpeter swans, and tundra swans stop off in the Mount Vernon area during their southern migration from Alaska; British Columbia, Canada; and the Yukon Territory to California. Today, the Johnson/DeBay Swan Reserve in northeast Mount Vernon, which is owned and managed by the Washington Department of Fish and Wildlife, is an ideal viewing location. (Courtesy of David Hock.)

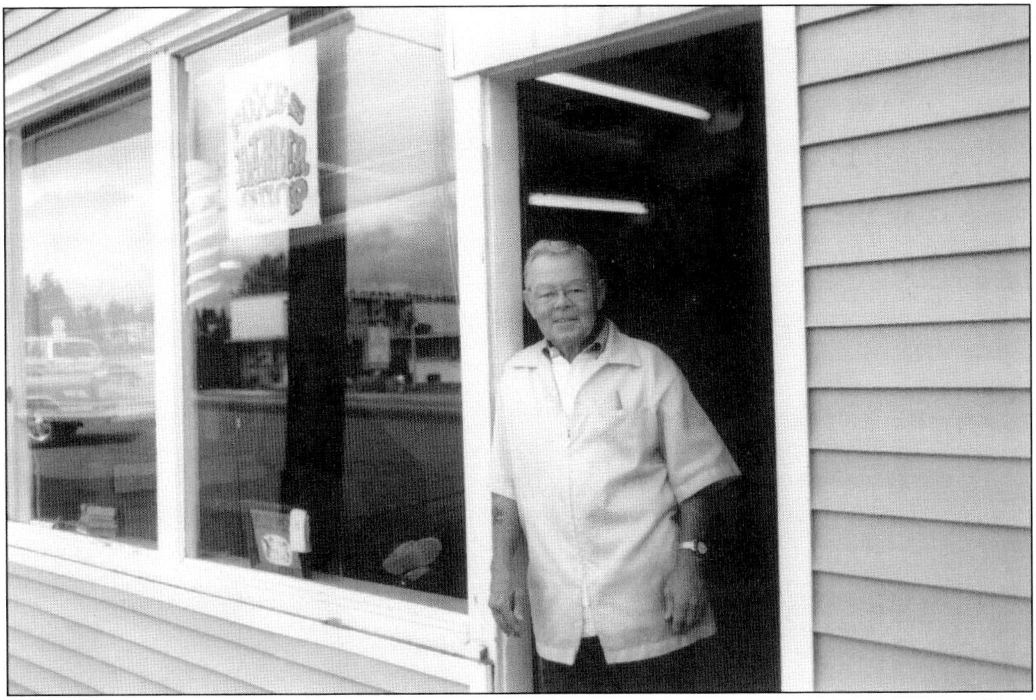

Pictured above is longtime Mount Vernon barber Roger Fox (January 30, 1933–June 26, 2004). He worked in the same west-side shop from October 1955 until his retirement on August 31, 2001 (also the date of this photograph). In addition to his barbering talents, he was a history buff and avid photograph collector. The Fox family had a history of barbering in Mount Vernon; in the 1920s, Roger's father, Orval, and grandfather William (both pictured below) ran the Paystreak barbershop at the north end of town. Orval started a new shop on the west side in 1947 and in 1955 was joined by Roger. Orval retired in 1982. (Both, courtesy of the Roger Fox collection.)

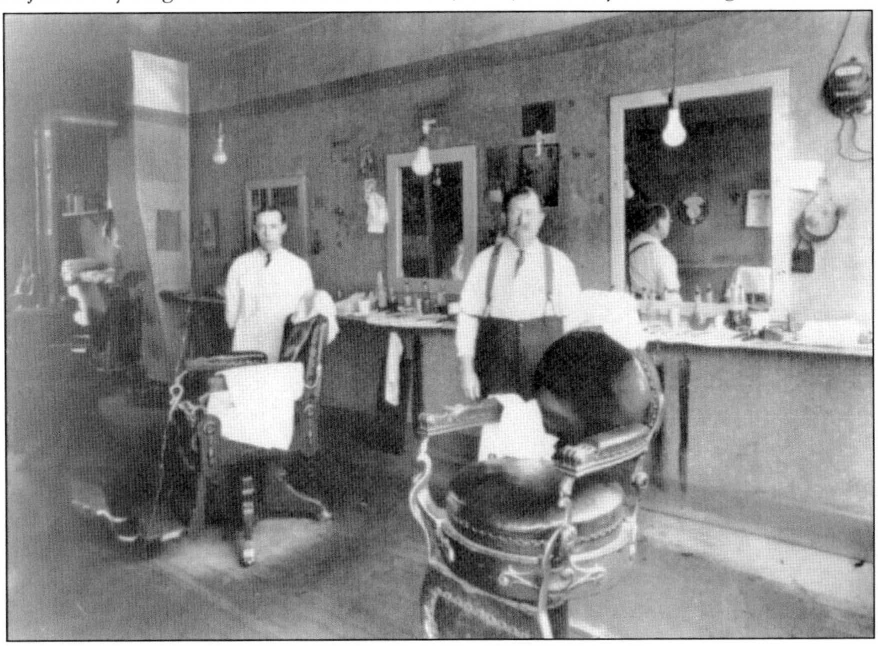

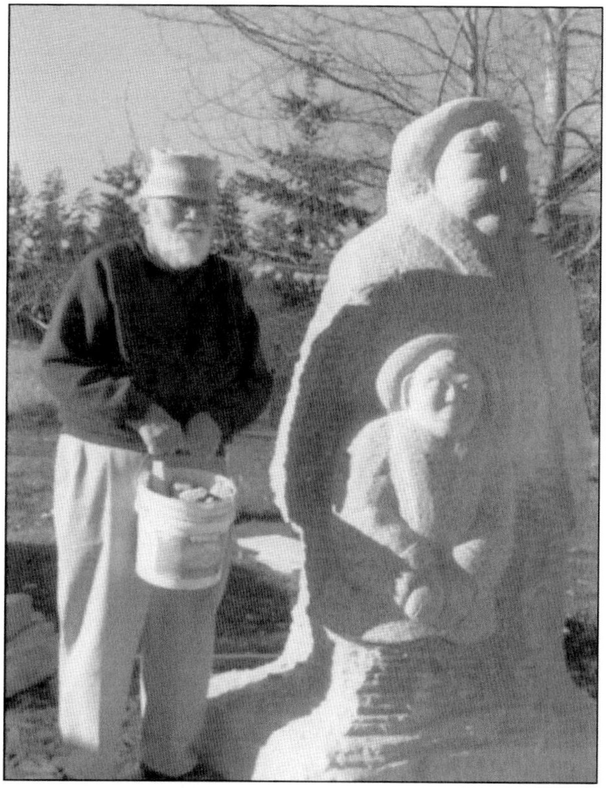

Richard "Dick" E. Fallis was born January 31, 1926, in Indiana and moved to Washington at the age of two. He had a passion for local history, which he put to use serving as historian for the Skagit County Pioneer Association, working at the Skagit County Historical Society & Museum as trustee and president, and coordinating the yearlong Skagit County Centennial Celebration in 1983. He wrote many books and countless historical newspaper columns. He also promoted and acquired funding for monuments, including the River Jam Memorial (below) and the Jasper Gates "Father of Mount Vernon" statue (right). Fallis is photographed standing beside the in-progress Gates sculpture, which was created by artist Tracy Powell and installed at First and Gates Streets in October 2005. Fallis died January 20, 2011. (Right, courtesy of Dan Royal; below, courtesy of Jessica Bylund.)

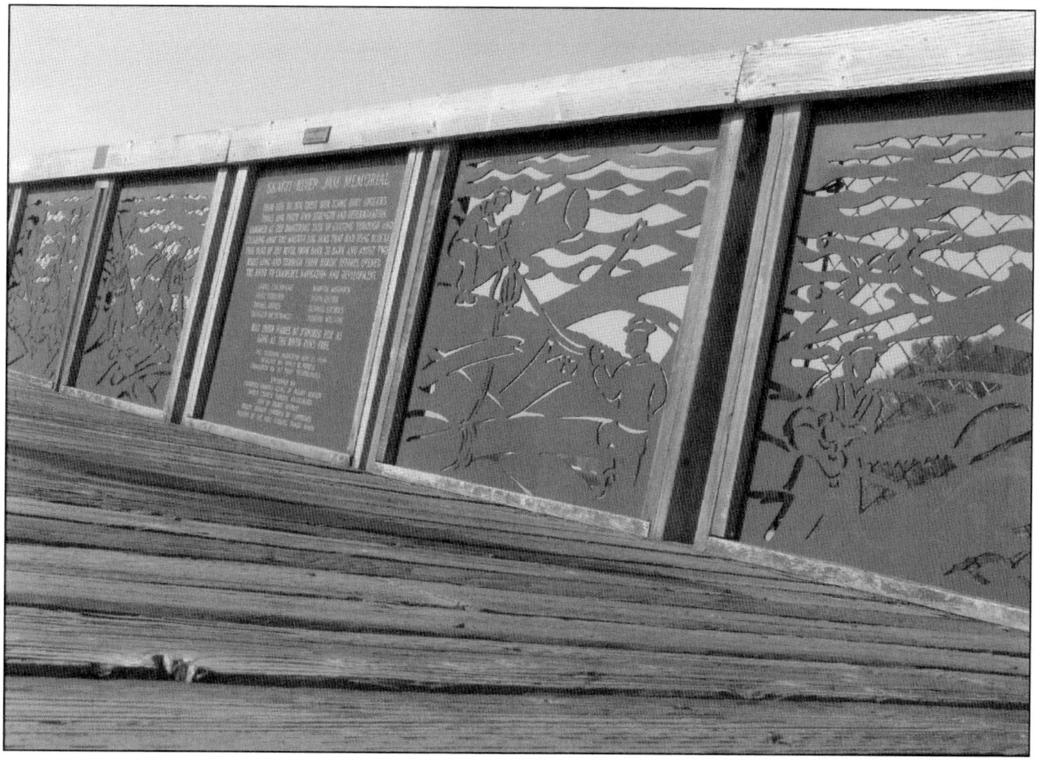

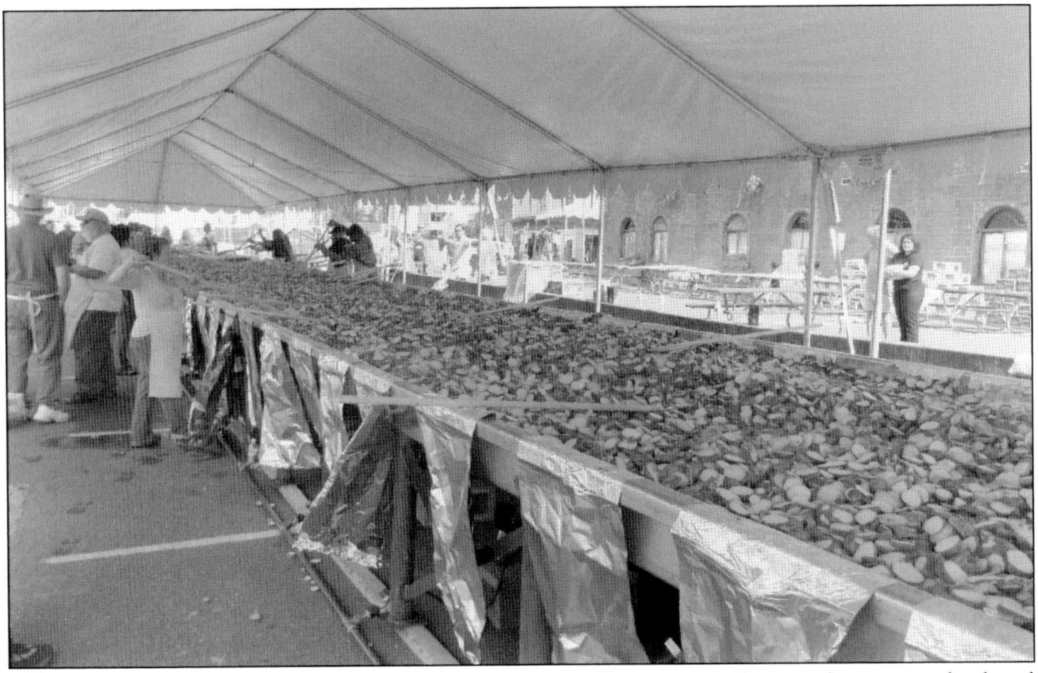

In 2011, Mount Vernon celebrated "Spirit of the Spud," a series of events honoring the local potato industry. On September 24, volunteers used 15,000 pounds of potatoes, 400 pounds of cheese, 100 pounds of onions, 10 pounds of garlic, and 200 gallons of cream to achieve a Guinness World Record for the largest potatoes au gratin. The event was sponsored by the Downtown Association but would not have been possible without the 10 local farmers, all members of the Skagit Red Potato Growers Association, who donated the potatoes. It was estimated that almost 150 volunteers helped throughout the process. When the dish was cooked and certified, it weighed 15,000 pounds, crushing the previous record of 8,400 pounds set in the Netherlands in July 2010. (Both, photograph by Konnie Knutzen McCutchin; courtesy of Ellen Gamson.)

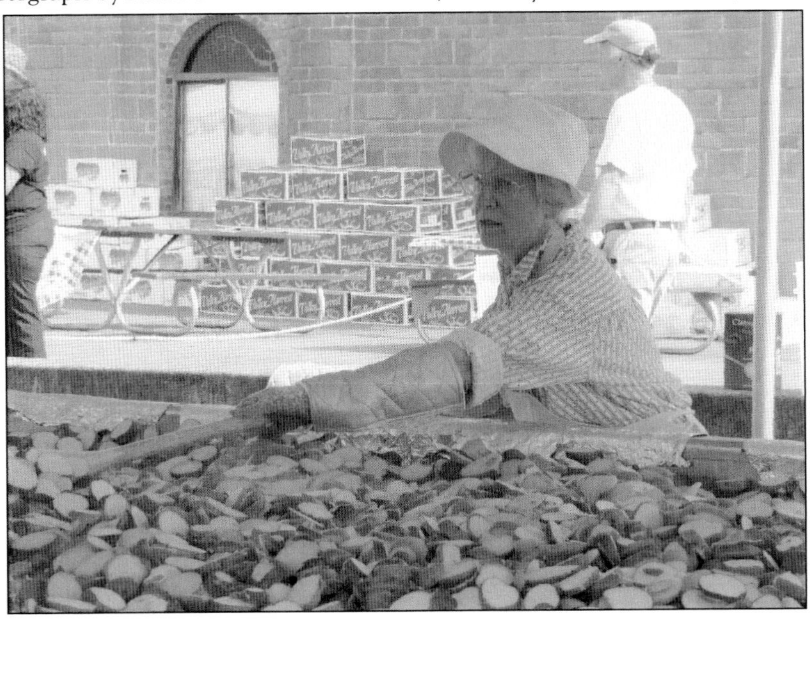

Bibliography

Barrett, Helen O'Brien, Anne Summers Carlson, and Margaret Willis, ed. *Skagit County Grows Up, 1917–1941.* Mount Vernon, WA: Skagit County Historical Society, 1983.

Bourasaw, Noel V., ed. *Skagit River Journal.* www.skagitriverjournal.com.

Caldbick, John. "Mount Vernon—Thumbnail History." HistoryLink.org.

Fallis, Dick. *Walking Tour of Historic Mount Vernon.* Mount Vernon, WA: Ashby Enterprises, 1986.

Fallis, Dick, and Dan Royal. *Bridgeside: Selected Stories and Dick Fallis Skagit County Historian.* La Conner, WA: Skagit County Historical Society, 2012.

Mount Vernon Argus, Mount Vernon, WA.

Pierce Jr., Ed, and Leigh W. Wilson. *"Let the Chimes Ring Out!": a History of Mount Vernon First Baptist Church, Mount Vernon, Washington, 1884–2007.* Mount Vernon, WA: self-published, 2007.

Skagit Valley Herald (formerly *Skagit News* [1884–1897], *Skagit Valley News* [1897–1913], and *Mount Vernon Herald* [1913–1956]), Mount Vernon, WA.

Solidaritet Lodge No. 396. *Scandinavians of the Skagit Area: a Collection of Historic Essays from Skagit County Pioneer Families of Scandinavian Descent.* Mount Vernon, WA: self-published, 2001.

Willis, Margaret, ed. *Chechacos All: The Pioneering of Skagit.* Mount Vernon, WA: Skagit County Historical Society, 1973.

Willis, Margaret, ed. *Skagit Settlers: Trials and Triumphs, 1890–1920.* Mount Vernon, WA: Skagit County Historical Society, 1975.

Wood, L. Stedem, ed. *The Skagit Collection: Our Proud Heritage in Pictures, Volume 1.* Mount Vernon, WA: *Skagit Valley Herald*, 2000.

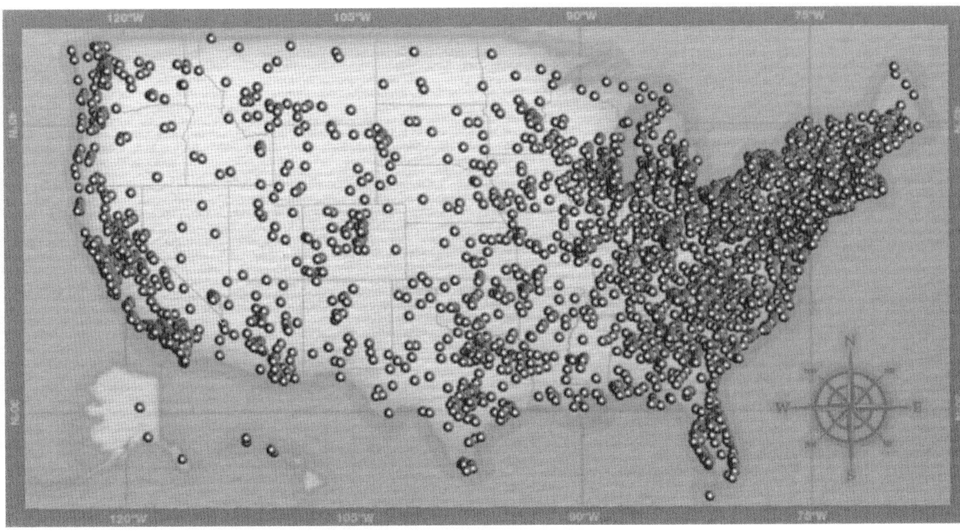